HOW TO HIRE
THE RIGHT CONTRACTOR

Other Books in The Homeowner's Library Series

Home Security
Floors, Walls & Ceilings
Creative Ideas for Household Storage
Preventive Home Maintenance
The Complete Book of Kitchen Design

THE HOMEOWNER'S LIBRARY

HOW TO HIRE THE RIGHT CONTRACTOR

Paul Bianchina
and the Editors of
Consumer Reports Books

CONSUMER REPORTS BOOKS
A Division of Consumers Union
Yonkers, New York

Copyright © 1991 by Paul Bianchina
Product Ratings (pages 137–154) copyright © 1991
by Consumers Union of U.S., Inc.
Published by Consumers Union of United States, Inc., Yonkers, New York 10703
All rights reserved, including the right of reproduction
in whole or in part in any form.

Library of Congress Cataloging-in-Publication Data
Bianchina, Paul.
How to hire the right contractor / by Paul Bianchina; and editors
of Consumer Reports Books.
p.   cm. — (The Homeowner's library)
ISBN 0-89043-396-8
1. Construction contracts—United States—Popular works.
2. Homeowners—Legal status, laws, etc.—United States—Popular
works.   3. Contractors—United States—Popular works.   I. Consumer
Reports Books.   II. Title.   III. Series.
KF902.Z9B53   1991
343.73′07869—dc20
[347.3037869]   91-23015
                                                                        CIP

Design by GDS/Jeffrey L. Ward
First printing, August 1991
Manufactured in the United States of America

How to Hire the Right Contractor is a Consumer Reports Book published by Consumers Union, the nonprofit organization that publishes *Consumer Reports,* the monthly magazine of test reports, product Ratings, and buying guidance. Established in 1936, Consumers Union is chartered under the Not-For-Profit Corporation Law of the State of New York.

The purposes of Consumers Union, as stated in its charter, are to provide consumers with information and counsel on consumer goods and services, to give information on all matters related to the expenditure of the family income, and to initiate and to cooperate with individual and group efforts seeking to create and maintain decent living standards.

Consumers Union derives its income solely from the sale of *Consumer Reports* and other publications. In addition, expenses of occasional public service efforts may be met, in part, by nonrestrictive, noncommercial contributions, grants, and fees. Consumers Union accepts no advertising or product samples and is not beholden in any way to any commercial interest. Its Ratings and reports are solely for the use of the readers of its publications. Neither the Ratings nor the reports nor any Consumers Union publications, including this book, may be used in advertising or for any commercial purpose. Consumers Union will take all steps open to it to prevent such uses of its materials, its name, or the name of *Consumer Reports.*

For Rose
—all that I've ever written
and all that I will ever write,
I owe to you.

Contents

Acknowledgments

I would like to express my gratitude to my agent, Deborah Schneider, for finding homes for my ideas; to Roz Siegel, for her excellent editorial advice and recommendations, as well as her patience; to Natalie Chapman, for all her hours and suggestions; and to all of the contractors, homeowners, and state agencies that put up with my questions and readily shared their experiences and advice.

And, as always, thank you, Rose—with your encouragement and quiet understanding, we've given birth to another one.

HOW TO HIRE
THE RIGHT CONTRACTOR

Introduction

If you're like most homeowners, there comes a time when you find it necessary to hire a building contractor. Whether for a minor repair, a major remodeling, or a complete new home, the bidding and selection process involved in finding the right person for the job can be a confusing and intimidating procedure—and mistakes along the way can have significant financial consequences.

Consumer protection groups across the country report that, year after year, of all the complaints they receive, the greatest number is against contractors. Discuss contractors with any group of homeowners and you'll find that almost all of them either have had a bad experience with a contractor or know of someone who did. As a result, many homeowners are reluctant to work—or even frightened about working—with a contractor.

Despite the bad press, the majority of contractors are both honest and competent, and most construction projects move from beginning to end with only minor, and essentially unavoidable, problems. In fact, adopting an overly aggressive or defensive attitude toward your contractor can set the working relationship off on the wrong foot from the very start.

Nevertheless, there are a number of contractors out there who are unqualified, unskilled, or downright unscrupulous. So how do you, a consumer with little or no building knowledge or experience, go about protecting yourself?

The Four Steps

There are four simple basic steps that can virtually assure that your rights will be respected, that your finances will be protected, that your job will run smoothly, and that you won't be victimized. Yet these steps—for the most part little more than simple common sense—are all too often ignored or overlooked. *How to Hire the Right Contractor* has been prepared, therefore, to guide you, with step-by-step detail, through a successful construction project.

1. Adopt a "Buyer Beware" Attitude. Don't assume that you are going to be taken advantage of, but don't passively accept everything you're told either. In selecting a contractor, take the time to verify licensing, bonding, credit, and a few other simple details; check references; go see a job or two if you can. Don't trust verbal agreements or your memory—get everything in writing to avoid misunderstandings later. This book will explain how to find and check on a contractor and provide information on all aspects of common construction contracts and change orders.

As the saying goes, "If it sounds too good to be true, it probably is." Beware of contractors who offer

- "big discounts" in exchange for huge cash deposits prior to the job
- cut-rate deals on materials "left over from other jobs"
- rebates if you use your home as a "model" to sell similar improvements to your neighbors
- to "trade" their labor for products or services you might have available through your own company

"Great deals" like these usually turn into great disasters. In Chapter 3 we'll take a look at many of the more common schemes and how to avoid them.

The fact that there are so many potential problems underscores the need to establish a comfortable relationship with a contractor. You want a person who will patiently clarify those things you don't understand. Honesty, fairness, and competence are certainly not unheard-of traits in a contractor, despite what your neighbors may have led you to believe.

2. Know What You Want. Before hiring a contractor, take the time to think over the work you would like to have done as clearly and thoroughly as possible. Your contractor can help you understand the construction process and make decisions about the costs and relative merits of what you have in mind, but first you need to have a strong idea of what you want.

We'll offer a number of tips on how to organize and understand your project before you begin, and explain how to properly prepare the project for bidding before calling in a contractor.

3. Clarify What You Want. In addition to knowing what you want, you have to know how to explain and clarify your ideas each step of the way. This will ensure that the contractor clarifies procedures and gives you the information you need to make decisions about materials and design.

By far the most common cause of problems between homeowners and builders is not dishonesty but simple misunderstanding. For instance, homeowners who don't know the proper terms often hesitate to appear foolish or annoying by asking too many questions. As a result, they have a hard time clearly explaining to the contractor exactly what they have in mind, and what they often end up with is not what they had visualized—or even contracted for.

Throughout this book we'll offer advice on how to communicate effectively with an architect, designer, or contractor, helping to ensure the results you want at a price that's fair and honest.

4. Know Your Rights and Obligations.   Signing any contract, large or small, is an important transaction. You need to know and understand what you're signing, what you can expect from your contractor, and what your contractor has a right to expect from you.

The goal of *How to Hire the Right Contractor* is to protect you, the consumer, and to simplify the entire construction process, making your next construction project an easier and more enjoyable undertaking. There's no magic and not even all that much mystery to building. During your next get-together with the neighbors, you can be the one boasting about how smoothly your job went!

1

Planning and Other Considerations

Long before the lumber trucks arrive and the hammers start to swing, long before you begin to call contractors and solicit bids, long before the bank or the building department or your nosy neighbors get involved, *you need to plan your construction project as completely as possible.*

That simple statement cannot be emphasized enough. The first and most essential key to a really successful project, and to a good working relationship with a qualified contractor at a fair price, is for you to know what you want. What goals will the construction accomplish? What problems will it solve? What do you want the finished rooms to look like? How much of the work do you intend to do yourself? Will you be acting as your own general contractor? How much will the work cost? Can you afford it?

These and many other questions need to be given a great deal of thought. You will be spending a large amount of money on construction; you will be experiencing weeks or even months of noise, dust, and inconvenience; and you will be altering how your house looks—how livable it is and what its resale value will be. Before doing all that, you should certainly know what you want and what to expect.

Know Thyself, Know Thy Project

With the exception of those situations in which homeowners encounter a contractor who has deliberately set out to rip them off (and those are actually pretty rare), and assuming the contractor they've hired is competent, it can safely be said that the vast majority of problems that arise between homeowners and contractors are the result of either *poor communication* or *poor planning*.

Typically, one of two things will happen on a job, and variations of these two themes have been played out between contractors and clients since the invention of the hammer:

• The homeowner did not fully plan out what he or she wanted—the size of the closets, the style of the moldings, the type of skylights, whatever—and was disappointed with the final results.

• The contractor did not make clear (or the homeowner didn't take the time to understand) what was included in the price, how long it would take to do the work, or how the finished job would look or function.

As in any relationship, communication is the key to understanding. Out of fairness to the honest contractor, it is important to say that a good portion of the responsibility for this communication falls to the homeowner. The contractor simply doesn't know what you want to have done until you tell him or her.

While a contractor can certainly suggest a number of layouts and materials and other options, you need to know the basics of what you want—and be able to communicate them to him or her. It is your best assurance of getting a finished product with which you're fully satisfied.

In addition to any notes you have made, pictures are often a great help in clarifying your thoughts and expressing your ideas to the contractor. While you're still in the early planning stages, start collecting magazines and books about construction and decorating. If you see something you like, whether it's a window grouping, a cabinet layout, or even just a molding or wallpaper pattern, catalog it for future reference. Most contractors won't mind flipping through your picture collection with you—it guides them in both the bidding process and the actual work, and helps ensure that their ideas for the project agree with yours.

Before your initial contact with the contractors, take the time to sit down with the family and discuss what you want *and* what you need (not always the same thing) from your new house or remodeling project. Establish as clearly as possible—down to room sizes, traffic patterns, furniture layouts, and decorating schemes—what you are trying to accomplish.

During these discussions, you will become aware of a number of questions that you really can't answer at this point: Is it possible to enlarge the bathroom? What types of materials are available for kitchen counters? These are things you should make note of so they can be discussed with the contractors. How well they answer these questions will be a good indicator of how experienced they are and how well you'll be able to work with them.

So, the first step in your quest for the perfect contractor begins in your own living room. Bring lots of paper and pencils, get the whole family involved, and note all the relevant comments and ideas.

While the following questions do not cover the whole spectrum of things to discuss and decide, they can help you focus your thoughts. They illustrate the types of things you should consider before diving into a project as big and expensive as this will be.

Why Are You Remodeling?

This always seems like a simple question, but when contractors pose it to homeowners, the answers they get range from "There's a waiting line for the toilet" to "The bedroom's too cramped" to "The kitchen is ugly." These answers are just not sufficient.

Kitchens, bathrooms, and room additions are the three most popular remodeling projects—and typically the three most expensive to accomplish. Taking them as examples, you can get an idea of the types of specific questions you should be asking yourself and discussing with your family.

For kitchen remodeling:

• Is the kitchen too small or too large? If so, how and where do you intend to enlarge or reduce it?
• What are the traffic patterns in and around the kitchen? How would you change them?
• Is the layout convenient for working? If not, why?
• Will this be a one-cook or two-cook kitchen?
• Do you need additional room for entertaining guests informally while cooking? Would direct access from the kitchen to the outside deck or patio be helpful?
• Are the appliances outdated, and if so, what new ones do you have in mind? Do you want to add a microwave or perhaps a trash compactor? How about specialty appliances, like a commercial cooktop, a deep fryer, a barbecue, or even a pizza oven?
• Is there enough counter space? Are you happy with the counter materials or would you like to change to something else? If so, what?
• Is the natural lighting sufficient? Is the artificial lighting sufficient? Would you like to add windows, skylights, or additional overhead or under-cabinet lighting?
• Do you need more storage space? Will additional cabinets be sufficient, or should you consider adding a pantry? Are there enough drawers?

- Do you like the current color scheme? How would you change it? Is there an overall look, feel, or style you'd like to achieve (early American, country, French, etc.)?
- Are you satisfied with the materials used on the floors and walls? How about the cabinets?
- What aspects of your existing kitchen would you like preserved in your new kitchen?

For bathroom remodeling:

- Who currently uses this bathroom—adults, kids, guests, everyone?
- Will the people who use this bathroom change in the near future (kids leaving home, an elderly adult coming to stay)?
- Will the bathroom be used by someone elderly or by someone with specific disabilities? (Bathrooms of this type require careful planning and often need special fixtures.)
- Is the existing bathroom too small? Where and why?
- Is the arrangement of the fixtures satisfactory or would you like to change them? (Give some careful thought to this one, since moving fixture locations, with the resulting alterations to the existing plumbing system, can really drive up the costs of remodeling.)
- Do you need to add additional fixtures, such as a bidet, a second sink, or a stall shower?
- Are you satisfied with the types of materials that are currently in your bathroom (ceramic tile, fiberglass, linoleum, etc.)? Do you want to use something different? If you want something different, do you know what and why?
- If you're adding a bathroom, who will be using it? Is a minimum size all right or do you envision a large, luxurious retreat?
- Where will the door to the new bathroom be? Will this entrance be conveniently located for the people who will be using it?

For room additions:

- Will the addition go out, up, or a combination of both?
- Will you need to incorporate some of the existing interior space into the new space (adding onto a kitchen to make it larger, for example)?
- What will the addition actually be used for? Will it be a single room or will the interior space be divided into several areas?
- What side of the house will you add on to? Which directions will the windows face? Is there potential for solar heating?
- How will you gain access to the addition from inside the existing house? If you are thinking of adding a second story, where will the stairs be located?

Have you allowed for the interior space in the existing house that will be lost to the staircase or the new hall that will serve the addition?

• How will the addition be finished on the outside? Will you use siding that matches the existing house or contrasts with it? How will you treat the roofing, windows, moldings, and other aspects of the addition that will need to either blend or contrast with the existing house?

• How about the interior? Will it blend with the existing house decor and architectural style or will it be something totally different? Two architectural and decorative styles can often be blended in one house with perfect harmony, but you need to plan for it.

Why Are You Building a New House?

This is another one of those seemingly simple questions that deserves some serious thought. On the one hand, it may be to your advantage to remodel your existing home rather than build a new one. On the other hand, it may not. This is a decision that has to be very carefully evaluated before you start a project of this magnitude.

• Do you need more—or less—interior space than what you have now?

• Do you want a house with a different architectural style from your present home? Do you want to live in a different neighborhood?

• Do you plan to build on a lot in the city or on acreage in the country? This can make a big difference in cost and type of construction.

• Will the house be one story or two? Who will prepare your plans—an architect, a designer, the contractor, or you? Will you be buying a set of stock plans? Remember: You need to have your plans ready before the builder can give you an estimate.

• What specific rooms do you need? Do you want a game room? How about a hobby room or a workshop? Will the kitchen be used just for quick meals or will you also be preparing elaborate meals on a regular basis? How many bedrooms and bathrooms do you want?

• Will any room need to be big enough to accommodate specific items or activities, such as a pool table, a grand piano, or a model train layout?

• Are there specific things that you want that will be a little out of the ordinary—a sauna, a spa, or even an elevator? The contractor should be told about such things early on.

• What type of heating system would you like? Do you want a fireplace or a wood stove? If so, in which rooms?

• What types of exterior materials do you envision? These will affect the cost significantly, so you'll want to consider if you want plywood siding, wood siding, brick, or stone veneer; wood, tile, or composition roofing; wood, vinyl, or aluminum windows; ornaments such as wrought-iron rails or wood shutters; and much more.

    • What about the interior? Stained moldings cost more than painted ones, oak costs more than fir, hardwood floors cost more than carpeting.

    • Will you expect the contractor to include other exterior work in his bid—such as decks, fences, or landscaping?

What Is Your "Wish List"?

As the ideas for your remodeling or your new home become clearer to you, you will find yourself discussing a number of specific details—things that you and your family would like to see included in the project. This is your "wish list," and it will develop as a result of conversations with your family and the contractors.

    Your wish list is important in helping you decide what you really want and need from the project, and what you can do without if necessary. Everyone starting a construction project has hundreds of great ideas, but it's rare to find anyone who can afford all of them. The trick is to know what to keep and what to eliminate.

    When you have become familiar with the requirements of the project, start a new list. Divide it into three columns; label the first one "Must Have," the second one "If Possible," and the third one "Can Do Without." Now, work back through the details of the project as they appear in your notes and put each item in one of these three categories.

    For example, if a new stove is essential in your kitchen renovation, it should go in the Must Have column. If the oversize pantry cabinet is less important, put it in the Can Do Without column.

    Be as honest with yourself as possible when making up this list. It is a surprisingly effective way of helping you establish and maintain realistic limits for your project while not eliminating anything you really want. If an unusual item—a brick pizza oven, for example—is really important to you, then by all means put it under Must Have. Remodeling or building a house and not getting the things you want and need is little better than not remodeling at all—you'll never be satisfied.

    Most people end up tossing out all of the items on the Can Do Without list. Plan on telling the contractor to include in the initial estimate only those items in the Must Have column. Then, if your budget allows, you can begin including selected items from the If Possible column.

What Is Your Budget?

This question doesn't always have a ready answer. How much thought have you given to the amount of money you can comfortably afford to devote to this project? Here are some budget-related questions to ask yourself:

    • Do you have any idea as to what remodeling—or a new house—currently costs, or will you be relying on the contractor for this information? The more you know about costs before you contact a builder, the less time you'll be wasting—both yours and his. Many homeowners, especially those who haven't had any construction work done for a while, are shocked to learn what construction work of any kind costs today. There's nothing worse than having your heart set on a new kitchen only to find out it will cost $18,000 instead of the $9,000 you'd had in mind.

• How much of the project will be financed through loans and how much will be cash?

• If you are planning on putting some of your own cash into the work— and practically every lender will want to see you commit some of your money before it will give you a loan—how much can you comfortably afford without leaving yourself too cash-poor to meet emergencies?

• Who will be lending the additional money for the work? Have you talked with anyone at a bank? Have you talked with mortgage brokers or other finance institutions? Do you have personal loan options open to you, relatives or friends on whom you can rely for all or part of the construction money you need?

• If you are financing all or part of the work, do you know what the payments will be? Added to any current payments and bills you might have, is the new payment affordable?

• How will you secure the loan—second mortgage, personal property, etc.?

While your contractor does not need to know the specific details of your financial situation, he or she has a right to feel confident that you can afford the work you are putting out for bid.

Many contractors have experienced the frustration of getting to the end of a job and finding out that the cost of the work has exceeded the homeowner's ability to pay. In such a case, either the job is left unfinished or the contractor is unable to get the final payments on time. Remember: A signed contract works in both directions—it protects the contractor as well as you. Your failure to pay opens you up to legal action just as readily as the contractor's failure to perform.

Will You Be Doing Any of the Work Yourself?

Many homeowners like to do some of the construction work themselves, whether it's to save money or just for the enjoyment of it. If you had planned on lending a hand around the job, you'll need to give some thought to what you'd like to do and how that will fit with the contractor's scheduling and normal sequence of work. Many homeowners limit their participation to demolition work, job site cleanup, trash hauling, and other labor-intensive chores. Some will do their own painting; others might undertake the plumbing or the electrical wiring.

• What tasks would you like to perform? Are you competent at them? (You'd hate to have an expensive room addition built and then do a bad paint job.)

• Do you have the necessary tools and equipment to do whatever work you're thinking of undertaking? If you want to do the cleanup and trash hauling, do you have a truck or a trailer to use? You may need to rent one, or you may be able to borrow one from friends or even the contractor.

• How flexible is your time? Most contractors won't mind if you want to do some of the work yourself, but they will insist that it be done in a timely fashion. They don't want to have a crew waiting around while you finish up the painting you said you would complete last weekend, or watch their drywall getting wet because you haven't completed the roofing you were supposed to do. Take a careful look at how much time you have available and when it can be scheduled before you commit to doing work on the project.

Costs and Options

Part of the confusion that occurs in the early planning stages of any construction project results from a lack of knowledge about how much things really cost. Take kitchen remodeling, for example. You know what you'd like to do with the room, and you've seen pictures and advertisements for cabinets, appliances, fixtures, and a hundred other materials that you think you'd like to use. But what can you afford? How much more would it cost to use a marble countertop instead of a tile one? Are oak cabinets more or less expensive than birch? It's tough to know what to ask for when you don't know what you can afford.

Early in your planning, take the time to start familiarizing yourself with costs. Take a weekend or two and visit some home centers, plumbing and electrical supply outlets, cabinet shops, carpet stores, and other building material outlets. All of this initial legwork on your part will be a tremendous help in learning about what items are available, what items appeal to you, and how much these things cost.

As you look, ask for brochures or photocopies of specification sheets. Make notes as to price, availability, and your opinions of the product. Be sure and note the name of the store you were in and, if possible, the name of the salesperson you spoke with. All of these notes could be important should you or your contractor need further information in the future.

Another way of determining cost is to specify exactly what you'd really like and have the contractors bid the job accordingly. Then, if you can't afford it, you can ask for suggestions on where to trim costs. This approach is fine to a point, but if you've specified a kitchen that the contractors bid out at $30,000 and all you have to spend is $10,000, it will be tough to do that much trimming. You will probably need to start over with a whole new plan, which means a lot of extra work for both you and the contractors.

Another approach is to make some comparisons, to see if you are continually selecting the most expensive materials. By looking at cost comparisons and then checking over your wish list, you can get a better idea of what's important to include and what isn't.

The following chart rates twenty common areas of labor and materials encountered on a construction project in order of relative cost—less expensive, mid-range, or more expensive. Average costs are included where possible. Remember: These costs are for *comparison only;* they can vary tremendously in different parts of the country, depending on availability of materials, shipping costs, labor rates, competition, markup, and a host of other considerations. Also, a few phone calls and a Saturday out doing some comparison shopping of your own will help a great deal in refining these averages.

New Construction

Few things can equal the excitement of having a new house built to your specifications. Unfortunately, few things can equal the cost either. If you are considering having a new house built, it will help to have an idea of the average costs for new construction across the country. These costs continue to show increases with the release of each quarter's government figures.

As of the third quarter of 1989, building a new home of "good" construction (denoting a home of mid-range quality and features, above a tract house and below a true custom) cost $59.25 per square foot. This figure reflects what a builder would charge a client and includes profit and overhead markups, as well as a two-car attached gargage but no basement; it also includes the basic excavation work needed to ready the site for building, but it does not include the cost of the land itself.

Upgrading to "better" quality construction—which usually describes the types of materials and extra amenities found in true custom homes—adds an average of 3 to 4 percent to the base cost. Building in a rural area can drop the cost by 4 to 6 percent.

If your home is over 2,500 square feet, the cost per square foot begins to drop by around 2 to 3 percent. It's actually more expensive per square foot to build a small house than it is to build a large one, since the expensive items—foundation, plumbing, wiring, heating, etc.—are amortized over a smaller area.

For example, building a 1,400-square-foot house of "good" construction—typically a three-bedroom, two-bath house with kitchen, living/dining room, an indoor laundry area, and an attached garage—would cost approximately $82,950 (1,400 × $59.25), excluding land. Upgrade the construction features and amenities into the

COST COMPARISON CHART

Item	Less Expensive	Mid-Range	More Expensive
Framing labor	*"Tract"-style framing*	*"Average" to "Good" construction*	*"Better" construction*
Cost:	$3.60–$3.80/ square foot	$4.00–$4.90/square foot	$5.20–$5.80 and up/square foot
Roof styles	*Gable end, manufactured truss*	*Gable end with intersections, site-framed*	*Hip and other styles, site-framed*
Cost:	Varies	Varies	Varies
Heating systems	*Standard gas; standard electric*	*Hot water; Condensing gas*	*Heat pump*
Cost:	$1,500 and up	$2,500 and up	$3,500 and up
Siding materials	*Composition sheets; plywood sheets*	*Waferboard or composition; lap or board*	*Cedar, redwood; special styles*
Cost:	Varies	Varies	Varies
Roofing materials	*Composition shingles*	*Cedar shingles or shakes*	*Metal or tile*
Cost:	$30–$70/100 square feet	$80–$120 and up/ 100 square feet	$120 and up/100 square feet
Windows	*Standard aluminum*	*Solid vinyl*	*Wood*
Cost:	$50 and up/4' x 4' slider	$100 and up/4' x 4' slider	$150 and up/4' x 4' slider
Skylights	*Fixed, acrylic, "bubble style"*	*Fixed, wood frame, flat glass*	*Operable, wood, flat glass*
Cost:	$75 and up	$150 and up	$225 and up
Wood decking, materials	*2x6 #2 fir*	*2x6 #2 cedar*	*2x6 #1 or clear cedar/redwood*
Cost:	$.30–$.50/linear foot	$.50–$1.00/linear foot	$2.00 and up/ linear foot
Interior wall covering	*Flat paint*	*Semi-gloss paint*	*Wallpaper*
Cost:	$.22 and up/square foot	$.24 and up/square foot	$1.00 and up/ square foot
Interior doors, materials	*Hollow core, flushed, hardboard*	*Paneled, hardwood*	*Solid core, flush, hardwood; paneled*
Cost:	$35–$50	$50–$80	$85 and up

Item	Less Expensive	Mid-Range	More Expensive
Trim—style	*Drywall-wrapped windows, standard casing patterns*	*Wood-wrapped windows, decorative casing patterns*	*Combination moldings; custom moldings*
Cost:	Varies	Varies	Varies
Trim— materials	*Vinyl or vinyl-wrapped wood*	*Solid softwood*	*Solid hardwood*
Cost:	Varies	Varies	Varies
Floor coverings	*Vinyl tile*	*Carpet; linoleum*	*Hardwood; ceramic tile*
Cost:	$5–$15/square yard	$15–$25 and up/ square yard	$40–$60 and up/ square yard
Cabinets	*Vinyl-faced particleboard*	*Modular alder or birch, flush or flat panel door*	*Custom oak, raised panel door; "European"*
Cost:	Varies	Varies	Varies
Counters— materials	*Plastic laminate*	*Ceramic tile*	*Wood; marble*
Cost:	$3 and up/square foot	$6 and up/square foot	$16–$60 and up/ square foot
Stairs	*Straight, softwood lumber*	*Carved or turned, hardwood*	*Curves, spirals, hardwood or metal*
Cost:	Varies	Varies	Varies
Sinks	*Thin-gauge stainless steel*	*Enameled steel; porcelain*	*Cast iron; heavy-gauge stainless steel*
Cost:	$30 and up	$75 and up	$125 and up
Bathtubs	*Enameled steel*	*Cast iron*	*Whirlpool*
Cost:	$100 and up	$200 and up	$800 and up
Toilets	*Economy or standard, reverse trap*	*Siphon jet*	*Siphon action "low-boy"*
Cost:	$80 and up	$125 and up	$300 and up
Job cleanup, labor	*Owner supplied*	*Contractor supplied, "broom clean"*	*Professional cleaning crew, new house*
Cost:	None	Part of contract price, average $50–$100 and up	$.10–$.15/square foot

"better" category and the price would be in the range of $85,450 to $86,300 (a 3 to 4 percent increase). Increase the house size to 1,500 square feet and the price climbs to $88,875 for "good" quality, and $91,550 to $92,400 for "better" quality. Looking at the numbers in this way, you can easily see the impact of increased square footage.

The following table shows where the money goes in constructing a new house. The second column shows the approximate percentage of the total cost of each item in the house, and the last column shows how that translates to dollars and cents, based on our 1,400-square-foot, $82,950 house. Of course, these are average costs, and they will vary with the exact amenities included and the materials selected.

Item	Percentage	Example Cost
Plans and permits	2	$1,659.00
Excavation	1½	1,244.25
Foundation	4½	3,732.75
Rough lumber	7½	6,221.25
Rough carpentry	8	6,636.00
Roofing	4	3,318.00
Windows	2	1,659.00
Siding	4	3,318.00
Fireplace, masonry	3	2,488.50
Plumbing, rough	7	5,806.50
Wiring, rough	3	2,488.50
Furnace and ductwork	5	4,147.50
Insulation	2½	2,073.75
Gypsum wallboard	5	4,147.50
Exterior painting	2	1,659.00
Gutters	1	829.50
Concrete flatwork	3	2,488.50
Doors and trim	4	3,318.00
Cabinets and counters	5	4,147.50
Finish hardware	1	829.50
Carpeting	3	2,488.50
Linoleum and tile	2	1,659.00
Interior painting	2½	2,073.75
Plumbing fixtures	2	1,659.00
Light fixtures	1½	1,244.25
Appliances	1½	1,244.25
Cleanup, final grading	½	414.75
Profit and overhead	12	9,954.00
Totals	100	$82,950.00

Regional Modifiers

The state in which you live, and even the area within that state, influences how much you will pay for construction work. Building in rural areas typically costs less than building in urban ones. Raw and finished materials may be in more abundant supply in some areas. The economy and the amount of competition may be stronger in one state than they are in another. All of these things and more can influence construction costs.

Prepare a Bid Sheet

Now that you've done your homework and filled several notepads with ideas, sketches, and questions, it's time to organize it all and get it ready for the contractors. There are two basic reasons for doing this. First, you want to be as detailed as you can about what you want, so the contractors can give you a realistic price. Leaving out several thousands of dollars worth of items because you haven't decided on them will make a big difference in the cost of the overall job. Once the job has begun, you certainly don't want to be forced to eliminate things or to use inferior grades of materials (or, worse yet, you don't want to be unable to finish the job) because the budget was inaccurate from the start.

The other major reason is that you want the bids from each contractor to be based on the same information so they'll be easy to compare. Remember: Bidding is a very competitive business for contractors. Since their estimates are usually free and can eat up dozens of hours, they can't afford to lose too many. They want to be able to compete fairly with the others you have bidding for the work. Telling one contractor that you want one thing and then telling the next contractor that you want something else can make a difference of hundreds or even thousands of dollars in a bid.

Preparing a bid sheet for the contractors is not difficult. It doesn't have to be formal, nor does it have to follow any particular order or layout. Some people type up whole lists of specifications, down to model numbers of appliances and colors of paint. Others simply make a rough, handwritten list. If an architect or designer prepared the plans for you, he or she will usually have a specification or "call-out" sheet that lists most of the pertinent information on doors, cabinets, windows, trim, and dozens of other details. It's vital that your bid sheet include all the items and that it's as complete as possible.

No matter how complete and well-thought-out your lists are, some contractors will probably suggest an alternate way of doing something. They may see something you've overlooked in your planning that needs to be included, or they may suggest a less expensive way of doing something. Perhaps they may know that something you wanted to do will not be structurally safe or will not satisfy the building codes. (Should this occur, it is wise to verify the opinion with your local building department before proceeding further.) Or, by virtue of experience with similar jobs, they may feel that you just won't be satisfied with the finished look or the layout you've planned and offer alternatives you might like better.

When this happens—and it almost always does—it's a good idea to take the extra time to seriously consider the suggestions. If you like them, then you can incorporate them into your plans and include them on a revised bid sheet. However, if other contractors have already bid the work and the changes you're considering are fairly major—at least from a cost standpoint—you'll want to provide them with details of the proposed changes so they can revise their bids.

Once again, what you're looking for are bids based on information that is very similar in nature and includes essentially the same labor and materials, within reason. You need to be able to look at the bids side by side to compare the prices, and you need a realistic estimate of what the job is going to cost so you can see if it will fit into your budget. You simply can't do that accurately if one contractor is bidding a 500-square-foot addition and the other is bidding one that's 600 square feet, or if one has included five windows and the other has figured on ten.

Acting as Your Own Contractor

A growing number of people each year are choosing to act as their own contractor, running personal construction jobs ranging from small remodels to large custom homes. Some do it for the experience and satisfaction of tackling a complex project and seeing it through to completion; others have had bad experiences with contractors and feel they would be better off in control of the project; but most do it for the money they can save—real or imagined.

As your own contractor, you will be faced with essentially the same duties as the general contractor, described in Chapter 2. You'll be dealing with all the government agencies, all the material suppliers, and all the subcontractors. You will also be responsible for knowing the building codes and constructing accordingly, and you will be inspected just as carefully as—perhaps even more so than—the professional builder.

When acting as your own general contractor, you will be working with any number of specialty contractors. For each one you hire, you'll need to follow the same general procedures, outlined in the following chapters—soliciting bids, checking credentials, getting written bids, and contracts—in order to ensure that your job runs smoothly and your rights are fully protected.

It's important to stress that being your own contractor is not as easy as it may seem, and it's definitely not for everyone. If you're undertaking a construction project for the satisfaction of doing it yourself, and you have the time and patience, that's fine. However, if you're doing it solely to save money, you may be surprised to find that it doesn't put as much cash back into your pocket as you might like—in fact, it may end up costing you more in the long run.

For example, many owner/builders must take time off from work—an hour here and there, sometimes an entire day or two at a time, occasionally an entire vacation—in order to run down materials or meet with subcontractors on the job to answer questions or solve problems. When you begin to take time away from your regular job, it could affect your income. It's a small step from that point to the project costing more than whatever money you might be saving by not having hired a builder in the first place.

Here are some important things to be aware of when considering acting as your own contractor:

1. It's Very Time-Consuming.   Depending on the size of the project you're undertaking, you may find yourself consumed by phone calls, material runs, and headaches and decisions of all shapes and sizes. The worst problem—and this is experienced by professional contractors and owner/builders alike—is scheduling. Construction jobs need to be scheduled in a particular sequence, with certain subcontractors and materials on the job before others. It's up to you to have those subs ready at the right time. That means having to contend with their schedules as well as your own—and as a homeowner who's probably never going to need their services again, you may find yourself at the end of the line, behind jobs for their regular customers.

2. You Assume All Responsibility.   If you order the wrong materials and hold up another subcontractor, you're responsible for the wasted time. If you incorrectly schedule something or someone, again you're responsible for the wasted time or wasted materials. You're also the one making all the decisions, and you're the person everyone turns to if there are problems.

3. Financing Difficulties.   Banks and mortgage companies are very reluctant to lend money to homeowners who want to do their own building. They have lost money in the past on loans to people with more good intentions than skill. Banks and mortgage companies don't want to be left with an unfinished house or with a remodeling project that was so poorly done it actually lowered the value of the home.

Banks work on statistics, and the statistics show that enough owner/builders have been poor risks in the past. Therefore, many banks now have policies against making construction loans to nonprofessional builders. This policy varies among areas and lending institutions, so check on financing before committing yourself to undertaking the construction.

One way around this dilemma is to use a contractor as a consultant. Many lenders are more willing to deal with a homeowner who has employed a contractor to oversee and coordinate the project, even though the homeowner may be doing most of the actual work. The lender may, however, require that the contractor be ultimately responsible for the completed project. Be sure and discuss this with your contractor after the bank's stipulations and restrictions have been determined. He or she can help you with the bank loan and serve as an adviser when you run into problems you can't solve. This will cut down on the amount of money you save, but it does provide a nice alternative to either hiring a contractor to do everything or plunging ahead and doing everything yourself.

2

What Is a Contractor?

The best place to start your search for the right contractor is to understand what contractors are, what they do, what licenses and qualifications they should have, and what type of contractor might be best for you. Knowing these things will make it easier for you to negotiate and deal with them as the job progresses.

Webster's New World Dictionary defines a contractor as "A person who contracts to supply certain materials or do certain work for a stipulated sum, especially one who does so in any of the building trades." This is a fair enough overview to start with. The two key terms in this definition are *contracts* and *stipulated sum,* which explain the essence of who and what a contractor is.

A contractor thus is a person who will visit your home, discuss with you the construction project you have in mind, and then prepare an estimate—the "stipulated sum"—of what that work will cost to complete. If you are agreeable, he or she will prepare and execute a contract with you that deals with all the particulars of your agreement. The contractor will then perform the work to your specifications and satisfaction for the agreed-upon sum.

In theory, this is how the process works; in reality, though, it is rarely that simple or that smooth. There are a number of pitfalls between the first meeting in your home and the driving of the final nail—ranging from simple mistakes and delays to financial disaster caused by the rare, truly dishonest contractor—and homeowners can fall into any or all of them. By following a few simple rules and doing a little homework, you can minimize these risks and keep your project as enjoyable and exciting as you had always hoped it would be.

The Business of Being a Contractor

Many people have misconceptions about what contractors do and where the money you pay them actually goes. While most successful contractors can certainly earn a decent living, a contracting business is not always the money-making machine that people often believe it is. Let's take a moment to learn how a contracting business operates.

Essentially, contractors generate income for themselves and for their firms in four different areas: labor, material markup, subcontractor markup, and profit percentage. There are, of course, many other avenues they can take to earn money. These include the construction and selling of new homes on their own or with financial backers, called speculative or "spec" building; the buying, renovating, and selling of older houses; and design and consultation work. But most of these other contractor services and endeavors will not concern you as a homeowner.

Labor

One of the contractor's primary methods of generating income is providing physical labor on your job (both the contractor's own labor and that of his employees). During the bidding process, the contractor estimates how much time the job will take; this is usually figured in number of days, but, for small jobs, it may be in the number of hours or, on really large projects, the number of weeks or months. The contractor will then apply the average daily labor rate—a rate that varies widely around the country—to the number of estimated days in order to arrive at a dollar figure for labor.

When bidding, some contractors will increase their rates on a per-hour basis, which lets them take into consideration the down time involved in moving equipment from one small job to another, or they may charge a minimum service call fee. For example, if you hire a plumber to fix a leaky faucet—usually about an hour's worth of labor—the plumber can easily spend more time than that just driving to and from your house.

On the other hand, a contractor may charge a lower rate when bidding on a per-week or per-month basis. It may well be worth it to him or her to charge less money to ensure that the equipment and crews are going to be guaranteed work on a large job for an extended period of time.

Specialty contractors often have other ways of calculating the cost of their labor for bidding purposes, all of which are essentially equal to whatever hourly or daily rate has been set. A mason or tile setter, for example, may have a labor rate of so much per square foot to install brick or ceramic tile. A concrete contractor may base the labor rate on so much per linear foot of foundation or per square foot of concrete slab. Carpet layers charge by the square yard, while a contractor specializing in finish work may base his or her rates on a "per box" price for setting cabinets or a per linear foot price for setting baseboards.

Material Markups

Contractors, by virtue of their license, can buy some materials at wholesale prices. They in turn mark these prices up—a little or a lot, depending on the job and the price they've paid—and resell the items to the consumer. In most instances, the amount of profit in the resale of materials is actually much less than people assume.

"Wholesale to the Public" signs abound at many large discount outlets, and wholesale pricing seems to have become one of the premier marketing gimmicks of the 1990s. When a consumer can buy a plumbing fixture for $100 at a discount outlet, the contractor certainly can't charge $175 for the same item. These outlets base their prices on large volume buying. Some items are even sold at a loss in order to get consumers into the stores, with the hope that they'll buy other, higher-markup items. In many cases, this discounted pricing is so low a contractor cannot compete with it. For instance, many plumbing and electrical fixtures and components are cheaper at the large volume home centers than at the wholesale house the contractor uses. In fact, many contractors shop these discount retail outlets for bargains just as you would.

Homeowners can actually buy lumber at the same price the contractor can in most instances, the only difference being that a contractor typically receives a small discount—1 or 2 percent—if the bill at the lumberyard is paid by the tenth of the following month. Let's assume that a contractor orders $3,000 worth of lumber for your room addition and arranges to have it delivered to your job site. Assuming the contractor has sufficient working capital—or assuming you have paid on time so that the money is available—he or she will pay his account by the tenth of the following month and receive a 2 percent discount. So, for driving to the lumberyard, selecting the materials and grades, examining the material if necessary, placing the order, arranging and scheduling the delivery, and, quite often, being there to receive and secure the delivery—all in all, about four hours' worth of time—the contractor will receive a mere $60 from the discount. This is certainly not what could be considered a large profit on the transaction.

Where the contractor's markup on materials becomes significant is with hard-to-find specialty items—things like special-purpose hardware items and fittings, available primarily through trade catalogs; some types of cabinet fittings and add-ons, such as specialty bins, trays, drawer inserts, and even fold-up ironing boards; custom fabrication, such as steel or brass fittings for framing; and decorative ceramics and hardwoods. The average consumer doesn't have ready access to many such items, or doesn't even know they exist. Markups of 100 percent and more are not uncommon, depending on the item, the wholesale price, and the difficulty and time involved in locating the item and placing the order.

On special-order items, shipping charges can sometimes be a money-maker for the less scrupulous contractor. This is one area to watch for and question, since the contractor has you somewhat over a barrel. If you don't have access to the distributor where the order is placed (and, in fact, you may have no idea who or where the distributor is), you'll have a hard time knowing what the freight charges actually are.

One woman was quoted $300 for some hand-painted ceramic knobs for her new kitchen cabinets. When the job was done, the bill for the knobs was over $500. She questioned the charge and the contractor tried to tell her that the manufacturer sent the knobs by special truck—and that the shipping charges were so high because the knobs were the only thing on the truck! Since she had not yet paid the contractor, she protested the amount and her contractor finally reduced the shipping charges to only $25.

When a contractor quotes you a price for a special-order item, be sure to ask if the price includes shipping. If it doesn't, ask for a binding estimate of what the shipping costs will be. If the contractor is unable to give you a binding estimate because, as sometimes happens, he or she won't know what the exact shipping costs are until the item actually arrives, ask that the distributor be called for an estimate. Get that estimate in writing, and do not pay more than 5–10 percent in excess of that original figure.

When inquiring about shipping, be sure you also ask whether the item is being shipped to the job or to the contractor's office. If it's going to the office, find out if there will be an additional charge to bring it to your home. One common gimmick is to quote you a specific figure for shipping the materials from the manufacturer or distributor, then later charge you an additional $25, $50, or even more as a "delivery charge," claiming that the price originally quoted was for delivery to the contractor's place of business, not to your job site.

Owner-supplied Materials.   There is, of course, an obvious temptation to avoid any markups, real or imagined, by purchasing and supplying the materials yourself. The subject of owner-supplied materials is something of a touchy one with many contractors and represents another area where understanding your contractor's business can make your relationship much smoother.

One contractor compares a client supplying his own materials on a construction job to a customer in a restaurant bringing his or her own eggs, asking the cook to fry them, and then complaining because the eggs aren't fresh. Viewed from the contractor's standpoint, this is not an altogether ludicrous analogy.

Suppose you intend to supply your own materials in order to save a few dollars. If the contractor installs your materials and there's something wrong with them—

they don't work, they're the wrong items, or there aren't enough of them—it becomes a subject of dispute over who is responsible and who is going to make any repairs. Also, if there are materials left over, you're responsible for returning them, just as you're responsible for getting more if the contractor runs short.

One homeowner actually had to leave work for four hours to get more tile for the tile setters who were working on his kitchen remodeling. They had run short because he had miscalculated the square footage needed, and they had no idea where to get more because he had bought the tiles on sale at a discount house. The workers on the job were at a complete standstill until more tiles arrived. Not only did the homeowner lose four hours of work, he was also charged for the time the tile setters were idle.

In most instances, it's best to leave the ordering and purchasing of the materials to the contractor—with your approval, of course. The savings in doing it yourself are usually not substantial, and the hassle seldom offsets the money you save and the potential liability you assume. A contractor who wants your job and is willing to cooperate with you on a fair estimate that fits your budget will usually work with you on discounting his or her markup on certain materials. The contractor may even supply certain items to you at cost.

Nevertheless, there are two possible exceptions to this rule of not supplying your own materials. If you are looking for areas where you can save a little bit of money on your construction project, you might consider an arrangement with your contractor in which you do the legwork involved in picking up the materials, transporting them to the job site, and unloading and stacking them where needed. A lot of the contractor's profit in material markup is eaten up by the cost of the labor involved in getting those materials from Point A to Point B, so if you have the time and the strong back to do some of this work yourself, you can pocket the savings. Remember: Work out *all* the details of any such arrangement in advance, and *get them in writing* as part of the contract specifications.

The other instance is where there are bid allowances—specific allowances in the bid for you to either select or actually purchase your own materials. Bid allowances are common with items of an ornamental nature, such as carpet, wallpaper, or light fixtures, where all the decisions about color and style are virtually impossible to determine prior to the job's starting (see Chapter 3). Once again, be certain that all the details concerning what you're responsible for supplying are worked out in advance and put in writing.

Subcontractor Markups

Like markups on materials, a contractor will often place a small markup percentage on the bids he receives from subcontractors. This is primarily to cover the time spent in phone calls, meetings, scheduling work and inspections, and the other time-consuming tasks of overseeing the subcontractors. The contractor may also be allowing money for unforeseen labor costs that he or she will need to cover for the subs, such as opening up a wall for the plumber and then patching it.

There is no way for you to find out how much the contractor has marked up the subs, if at all. You can ask, but you can't really verify what you're told. You can ask the subcontractor directly, if you know the person, but it's doubtful you'll get an answer. Remember: The sub may work for you only once, while he or she may do hundreds of jobs for the contractor; the sub is certainly not going to do anything to jeopardize his regular working relationship.

The best protection you have against excessive subcontractor markup is competitive bidding for the whole job. A new house or a remodeling project will have a certain amount of wiring and plumbing in it. While it's true that there are several ways of doing things, especially in remodeling, each contractor should allow relatively the same amount for each of the subcontractor operations. If you like a particular contractor and his or her bid is competitive with the others you've received, you can feel confident that the subcontractors (and the materials, for that matter) have not been excessively marked up.

Profit Percentage

This is truly the bottom line, the place where contractors place a certain figure for profit and overhead. They may use a single percentage to cover both, a separate percentage for each, or a single percentage for profit alone if their overhead has already been figured into their costs throughout.

Typically, contractors use a bidding sheet when preparing an estimate. They calculate all the labor and materials that will go into the job, plus the cost of all the subcontractors. They subtotal the bid at this point to arrive at the actual cost of doing the job—known as the "hard costs." They then multiply this by their standard percentage—it could be as little as 5 or 6 percent or as much as 30 percent or more—to arrive at their allowance for profit and overhead.

In a thriving market the percentage may be higher than it would be in areas where the building market is slow. Profit margins also tend to drop in areas where there is a lot of competition for work because of an abundance of local builders or the desirability of a particular job. On the other hand, those contractors who specialize in a trade that few other people in the area can do, or who are popular and in demand for whatever reason, may command higher profit margins.

Regional factors can also influence profit margins. Certain cities and states are more expensive to do business in than others, with higher taxes or insurance, higher licensing fees, or additional regulations that eat up additional time or money. While these items are more in the category of the contractor's overhead expenses, they can affect the percentage allowed for profit as well.

Bear in mind that every business, no matter what type, needs to make a profit in order to stay in business—a contractor is no different. As such, his or her profit needs to be viewed differently from the labor he or she charges. The labor charge is essentially a personal wage—paid in order to have an income for personal expenses—while the profit is paid to his or her company so it can grow and meet its obligations.

National Average Markups.   There are a variety of things that affect a contractor's markup on materials and profit, including the size of the marketing area, the strength of the building market, and the amount of available competition. A recent national survey of contractors showed the following average markups:

Profit: 8 percent
Overhead: 10 percent
Contingency: 2 percent (not all contractors add a contingency fee)

A Look at Where the Money Goes

A contracting business is really no different from any other business. There are wages, there are expenses, and, if all goes right, there is profit.

Take, for example, a small, independently owned, local shoe store where you go to purchase a $40 pair of shoes. Of the $40 you pay for the shoes, $25 goes to the wholesaler who supplied the shoes; $10 goes to the proprietor's overhead (rent, the cost of maintaining the store, paying the utilities, advertising, etc.); $2 goes toward employee salaries; $2 becomes profit for the store and goes into the bank for emergencies and lean times; and the final $1—gross, before paying personal income taxes—is the owner's salary. Different stores would have different breakdowns and different expenses, but the principle is the same.

The store owner's expenses continue whether there is one customer a day or a hundred. The employees are paid to show the shoes whether or not they make a sale. If a high-salaried employee quits and is replaced with a low-salaried person, the increase in the bottom line is the owner's to keep—you probably won't see it reflected in a drop in the selling price of the shoes.

A contractor's business is essentially the same. A typical job might break down as follows: about 50 percent of the money you pay the contractor goes to purchase the materials; 15 percent goes to subcontractors; 15 percent is paid out in salaries; 5 percent goes to business fees and taxes, insurance, licensing and registration fees, and bond payments; 5 percent is eaten up by other overhead (office space and equipment, utilities, fuel, payments on trucks and equipment, tool replacement and maintenance); and the last 10 percent is pretax profit for the business.

So, when you hire a contractor, you are paying for a number of things, including estimating time, design and consultation time, tools, bonds and licenses, all miscellaneous supplies, all of the various taxes that any small business must pay these days, and, above all, the contractor's knowledge and experience in helping develop, oversee, and build a structure that works.

These expenses are part of the cost of doing business for anyone, and what you're paying for here is exactly the same as what you're paying for when you buy that pair of shoes, or when you purchase any goods or services. However, there are two crucial differences between the contractor and the shoe store owner; understanding the differences can make your dealings with the contractor much more pleasant for both of you.

Estimating

A contractor's estimate is just that, an estimate—a guess, to put it more bluntly, albeit an educated one. Especially in a remodeling situation, the contractor is making a number of guesses as to how long each phase of the job will take to complete—what will be found when a wall is opened, what will happen to the plumbing service or the electrical panel, how long it will take to frame a new roof line that will match the old one.

More experienced contractors may be more accurate in their time estimates (an advantage gained when dealing with experienced people). However, there is always the possibility that a task for which two days have been allocated may take four, while something the contractor figured would take four days takes only two. The contractor can only hope that the time allotted for each task evens out in the end and the job is completed within the total estimated time. If you're working from a set contract price, you will not get a credit if the contractor saves two days somewhere, just as you won't be charged extra if two days are lost somewhere else.

Negotiating

Negotiating for the price of the work to be done can become both a sore point and a potential problem. You would never think of asking the shoe-store owner if he or she would take $35 for that $40 pair of shoes you're interested in, or asking the restaurant owner to knock $10 off the dinner bill if you supply your own dishes and silverware. The contractor's bid reflects his or her estimated costs for the work. To offer less is going to do nothing but cause hard feelings.

That's not to say that a contractor won't work with you to give you a price you can afford. If you tell the contractor that your budget for the addition is $23,000 and you can't afford to go any higher, he or she can usually suggest minor changes or deletions that will save you money. Striking a business relationship based on mutual trust and honesty will go a long, long way toward keeping the job running smoothly and everyone smiling.

What Type of Contractor Do You Need?

The word *contractor* is a catch-all for just about anyone doing business in the construction industry, and many people do not realize that there are actually dozens of types of contractors. Knowing the kind of contractor you'll need for your project can simplify the bidding, ensure the services of an experienced person, and often save you money in the long run.

General Contractors

Mention "contractor" and the type of person that springs to most people's minds is the general contractor. The most visible person on most job sites, the general contractor is the person everyone turns to with problems or when decisions need to be made. He or she is the person with whom you will most likely deal on any large construction project.

Different states have different names and job descriptions for the general contractor, but essentially this is the person who is hired to maintain overall control of a construction project, large or small. In most states, general contractors handle both residential and commercial jobs. They do not manage large public works projects; these fall to public works or general engineering contractors.

While a general contractor typically has one or more specialty—framing, finish carpentry, masonry, or whatever—he or she is fairly well versed in all or most aspects of the construction trades. Although the general contractor's firm may perform only one or two physical tasks on the job (such as erecting the rough framework of a house or installing the trim), he or she will oversee all the materials and labor going into the project and will act as the final authority for the whole job. If your construction project requires the use of three or more building trades—carpentry, plumbing, and electrical wiring, for example—then you will be best off with the services of a general contractor.

The duties of the general contractor are many and varied. What follows is the typical sequence of tasks that the general contractor performs:

Meets with Clients. The process begins with a meeting between you and the general contractor to go over the plans and specifications for the home, usually taking anywhere from two to ten or more planning sessions before contracts are signed and work starts. One of the general contractor's prime duties on any project is to be the liaison between the homeowner and the architect, or designer if there is one, and all of the subcontractors on the job. You may have fifty or sixty different people working on your new house during the course of its construction, but you'll probably meet or talk with only a handful.

Prepares the Bid.   Following the meetings, during which a number of details will be ironed out about what the client wants and how the house is to be built, the contractor will prepare a bid. The bid will detail what tasks he or she will be responsible for performing, what materials will be supplied, and what the price will be. During the bidding process, the contractor will check with dozens of sources for prices on materials and subcontractor labor and will be in touch with the homeowner or the architect to clarify details or suggest changes in the specifications.

Executes the Contract.   If the bid is accepted, the general contractor will next prepare a contract that details his or her responsibilities, as well as the client's. The contract will also specify the final agreed-upon price, set the start and completion dates, set the payment schedules, and clarify a number of other details. If the state so requires, the contractor will also provide the client with specific written or verbal information pertaining to licensing, liens, and other legal matters.

Constructs the House.   With the details agreed upon and the contract signed, work can begin on the house. During the course of construction, the general contractor will order materials, arrange for subcontractors, schedule the flow of work in the proper sequence, answer questions, oversee changes, and act as liaison between the client and subcontractors and material suppliers. The general contractor will also oversee any of his or her own employees on the job and will often personally perform some of the construction labor.

Scheduling is the general contractor's other prime function during construction. In order to ensure that the house is completed on time and on budget, there is a series of phone calls to make and meetings to have: examining, selecting, and ordering materials; contacting the proper subcontractors and coordinating their schedules with the project's; dealing with city and county building officials and scheduling inspections at the proper times; contacting the bank and scheduling its inspections at the proper times, as well as arranging for the timely payment of all bills; and dealing with changes, back-orders, and decisions of all shapes and sizes.

Finalizes the Details.   As the job comes to a close, the general contractor will oversee each of the final details, arrange for corrections of anything unsatisfactory (to the contractor or the client), and arrange for final inspections by the bank and the building department in order to obtain all necessary approvals and releases prior to occupancy.

Specialty Contractors

Unlike general contractors, specialty contractors deal with only one trade. The specialty contractor may be hired by the general contractor on a project—hence the term *subcontractor*—or may contract directly with the homeowner. In some instances, the specialty contractor may also hire other subcontractors. During the construction of a swimming pool, for example, the swimming pool contractor may need to arrange for a plumber, an electrician, or perhaps even a carpenter to build some decks.

If you have a project that requires only one trade—for example, having a furnace replaced or having a new roof put on—you will probably want to deal directly with a specialty contractor. Like the general contractor, the specialty contractor will meet with you to discuss details, prepare a bid, execute a contract, and do all the necessary scheduling and follow-up work to get your project completed.

Specialty contractors conduct business in much the same manner as general contractors, with basically the same methods of consultation and estimating. They can also help you with your building plans and ideas, and can arrange for permits if necessary. However, unlike general contractors, they usually won't help with variances, zoning changes, and other land-use situations.

Architects and Designers

If your construction project is a large one—a new home, a complex addition or remodeling, or extensive exterior decks and landscaping—you may consider the services of an architect or a designer. One of the most difficult phases of any construction project is the planning and preparation—getting your ideas to come together into some sort of cohesive, workable plan—and the services of a person skilled in this field is often invaluable. While these services will add to the overall cost of getting the project done, you'll find that money is saved in the long run through the elimination of design errors—errors that can lead to wasted materials and construction time, as well as frustration.

Many of the precautions and guidelines suggested for dealing with a contractor also apply to dealing with architects and designers. A great deal of care needs to go into selecting the right person, checking references, and establishing the guidelines for what will be done and how much will be charged. As with the contractor, the relationship you establish with an architect or designer needs to be one of trust and respect, and you need to find a person who will guide you through the design process with a firm hand while still allowing you the creative freedom to design the type of house you desire.

Architects

There are two primary differences between an architect and a designer: the architect has considerably more formal design and technological education than a designer, and the architect is also state-licensed, an important advantage in some instances.

Architects go through a grueling five to seven years of college, usually finishing with a bachelor's or master's degree in architecture, depending on the program. College courses place a strong emphasis on progressively more complex design problems, ranging from simple vacation homes through large municipal buildings and even entire cities. There are courses in art and history with an emphasis on architectural design, as well as a number of classes in the technical side of building, including geology, physics, and structural engineering.

Following college, the newly graduated architect must serve an internship of some sort before becoming licensed. Internship programs vary from state to state, with some offering formal apprenticeship programs and others relying on practical experience gained through employment with an architectural firm. The apprenticeship is typically three years and must be verified by one or more supervising architects.

Following this internship, the architect is qualified to apply for licensing. State tests are administered once or twice a year and involve a difficult multiday, multipart examination stressing design and architectural technology. Once licensed, the architect is free to operate a business and seek clients, although some choose to continue as an employee of an established firm.

The standard image of an architect is usually someone like Frank Lloyd Wright—a person working on towering, immortal buildings and out of the reach of the average consumer. But most architects—Wright included—design custom, single-family homes, and some even specialize in this area. Many will also take on remodeling projects, although some handle only large renovations. There are also landscape architects who specialize in exterior designs for both residential and commercial buildings.

The advantages to working with an architect are usually most apparent in home designs or remodeling projects in which the site or the design requires complex structural considerations, such as difficult terrain, combinations of a number of dissimilar materials, or the use of large, open, unsupported areas. Because of the architect's much greater exposure to structural engineering during college and internship, he or she is usually much better qualified to make the required calculations and specifications than a designer.

Also, the architect's state license allows him or her to certify the structural integrity of the buildings he or she designs. In some instances, certification is required by state or local building officials prior to the issuance of building permits or even insurance. As a result, the architect assumes a much greater risk than the designer, as certification may hold him or her legally liable in the event of a structural defect in the building. In this litigious age, many architects carry what amounts to malpractice insurance to protect them in the event of a lawsuit, and this fairly hefty additional overhead is usually reflected in their fees.

Fee structures for architects vary, based on the complexity of the project and the size and type of the firm you're dealing with. The architect may charge by the hour or by the square foot for the size of the structure being designed, or there may be a flat fee for seeing the project through from consultation to finished drawings. Ongoing consultations and on-site supervision may or may not be included in the flat fee or the square-foot prices; these are points you'll want to agree upon from the outset.

Once you retain an architect for your project, you will probably be asked to sign a standard American Institute of Architects (AIA) contract, which spells out the details of your agreement—price, services, completion dates, and the liabilities and responsibilities assumed by both parties. Like any contract, read it over carefully and be certain you understand all of its ramifications before signing. This contract is with the architect only, and, unless he or she is hiring the contractors directly, will apply only to design and drawing services. You will have a separate contract with your contractor.

Designers

There are actually a variety of different types of designers, all with varying amounts of training and education. Many colleges offer fairly comprehensive two-year architectural design programs, with curricula that include drafting and design classes, basic architectural theory, and a number of practical building and basic engineering classes.

College is not a prerequisite for becoming a house designer—in fact, there really are no requirements. Some designers have an architectural degree but have never become state-licensed. Others learn the trade by working for architectural firms as draftspersons or by working for other designers. Still others come to the profession with a hands-on building background, having first been carpenters or contractors.

Because designers are not state-licensed, they cannot officially stamp and certify the structural calculations on the buildings they design. If you have a house or other structure that the building inspectors or other governmental agencies require be certified, you will need to hire an architect or a structural engineer to review the plans and make the necessary certifications.

Designers may charge by the hour, by the square foot, or by flat fee, but they are typically less expensive than architects—sometimes considerably so. Most can also offer you consultation and supervision services, and, again, there may be extra charges for this.

Designers, like architects, are skilled and trained to develop the ideas for the project, transfer them onto paper for your examination and approval, do the necessary structural calculations, and prepare the final set of drawings for your use in obtaining financing, securing building permits, and guiding the contractor. Most do not perform any of the actual construction work.

There are, however, a number of contractors who also do design work. They may have portfolios of house plans to choose from if you're interested in a new home, or they may have an in-house designer or draftsperson to work with you on your remodeling plans. For smaller projects, most competent contractors can prepare a basic set of plans for a project on which they're working and can also offer some design suggestions.

Duties and Responsibilities

Whether you need an architect or a designer—or neither—really depends on the size and complexity of the project and your budget. The differences between the two are explained below, but essentially both will help you design the living space, and both can perform a number of other important services over the life of the construction project. Their duties include:

Initial Design Consultations.   This involves one or more meetings and is intended to allow you and the architect or designer to get to know each other and review the proposed project. These meetings allow you to get a sense of the person's ideas, working methods, and fees, as well as whether or not you feel you can work together. It also gives the architect or designer the opportunity to see what you have in mind and determine if it's the type of project in which he or she is interested and competent to handle. Typically, the first meeting lasts an hour or so, and there is no charge.

Subsequent Design Consultations.   If, after your first meeting, you both decide to proceed with the project, the architect or designer will begin by asking a series of detailed questions, covering everything from how big you'd like the house to be to how much entertaining you do and how much you would like to spend. This provides a better understanding of your needs, so that he or she can design a structure that meets them. There may be several of these meetings before any drawing begins.

Initial Site Visits.   The architect or designer will visit your home or your building site to look at exactly what there is to work with. If you are remodeling an existing home, detailed measurements, sketches, and perhaps even photographs will be taken. If it's a new residence, he or she will examine the lot, look for views and obstructions, study the slope and contour of the land, and attempt to visualize where the house will stand, which way it will face, and how it can best be adapted to the lot. If the lot is large, irregular, or steeply contoured, the architect or designer may also arrange for a surveyor to come out and do a site survey or a topographical map.

Preliminary Sketches. As the design process proceeds, the architect or designer will begin generating rough sketches of floor plans (how the home's interior space is arranged, drawn as if viewed from above) and elevations (drawings of what the exterior will look like). For a new house, a site plan will also probably be prepared, showing how and where the home will sit on the lot. Dozens of preliminary sketches—many of which you'll probably never see—may be generated before arriving at a design you're both happy with.

Final Drawings. The last step in the design process is to develop a full set of working drawings. These drawings, which are highly detailed, will show where the house will sit, the exterior appearance of the house from all sides, the floor plan, foundation and roof details, framing and trim details, electrical layouts, and more. These plans are intended to be used, first, for the bank and the bank's appraisers in determining the value of the work; second, by the architectural review committee in your neighborhood, if there is one, for determining if the house meets the area's standards for appearance and materials; third, for the building department to review in making their evaluation of the home's structural integrity and compliance with codes prior to issuing a building permit; and finally, by the contractor and the subcontractors in bidding the work and building the house.

Ongoing Consultations. One of the other helpful services provided by designers and architects is ongoing consultation and supervision of the construction project as it progresses. How much or how little ongoing supervision you get is really up to you—there is, of course, a fee for it. Some architects may charge a flat fee or a percentage of the job's final cost for as much consultation and supervision as you and your contractor require, or you may be charged by the hour for the actual time spent.

Most architects and designers have enough pride and interest in their designs to want to see how they are progressing and to correct things that are wrong, either through their own omissions or builder error. Typically, at least one or two site visits will be made at no charge—this is, after all, a visible statement of their skill and creativity—and most architects and designers will answer occasional phone questions free of charge.

Employing the services of the architect or designer on an ongoing basis, even if only for infrequent consultations, is usually a wise investment. You have spent a good deal of money on designing and constructing the project, so a little extra expense to ensure its being built correctly is a good idea.

3

The Bidding and Selection Process

The selection of the right contractor is probably the single most important aspect of the entire construction process, and yet it is the one given the least amount of time and thought by most homeowners. The average person agonizes over designs and budgets and construction details large and small but is usually unwilling to spend more than a few hours picking the one person to whom all the money and all the dreams will be entrusted.

The bidding and selection process has several phases. Each one is equally important and relates to the ones before and to those that follow.

1. *Know what you want:* Determine, in as much detail as you can, what you want to have done and how much you can afford.

2. *Write it down:* Prepare a written bid sheet for the contractors to refer to when they come out.

3. *Solicit bids:* Obtain the names of contractors who do the type of work you're interested in having done and arrange for bids.

4. *Get verifications:* Verify who the contractor is and that all his business papers are in order.

5. *Talk with others:* Call former clients to find out about the builder. Arrange to see some past work if you can.

6. *Compare the bids:* When all the bids have come in, compare them with one another for price and completeness.

7. *Select the contractor:* Make a selection based on compatibility, competence, and price.

This procedure may at times seem tedious, but it has a number of very positive benefits, and a little extra effort here can really save some headaches down the road.

Licensing and Regulation

Prior to actively soliciting bids from the contractors, it is important to know what licensing, if any, your state, county, or city requires contractors to have. Once you know what the regulatory agencies require of the contractor, you'll have a much better idea of the questions you'll need to ask and what information you'll need to verify.

The one consistent thing about the licensing and regulation of contractors is that there is absolutely no consistency. In a society that takes great pains to examine the person who will cut your hair or that regulates the minute details of a transaction to buy a used car from an auto dealer, all too little attention is paid to the person who will assemble that expensive and extremely intricate collection of boards and wires and pipes that make up a house. It is disconcerting to realize that, in a surprising number of states, the person who just bought his or her first hammer yesterday could be building you a $300,000 custom home today—with little or no regard for your safety or your financial protection.

Some states—California and Florida, for example—do have tough licensing laws pertaining to contractors and equally tough laws dealing with the contracts they prepare. Other states may only require proof of financial stability or basic experience, while still others have no requirements or restrictions whatsoever. To further muddy the waters, about a quarter of the states have no statewide contractor regulation, leaving that task instead to the individual cities and/or counties where the construction work is being performed.

There is a similar lack of consistency with regard to bids and contracts. Some states spell out specific items that a construction contract must contain, while others don't even require contracts. One state may insist that you be informed about construction liens, while another may have no provisions on their books that even deal with liens. Some regulate what a bid must contain, others limit the amount of deposit a contractor can collect, and still others turn a blind eye to everything and leave it to the veracity of the contractor and the common sense of the homeowner to police the industry.

Appendix A is a state-by-state listing that details the basics of what each state does—or in most cases doesn't—require of the contractors doing business there. While this should serve as an introductory guide, don't let it stop you from doing your own homework. Bonds and insurance can vary between contractors, depending on what they do and the volume of work they handle. Similarly, the laws governing contractors change from county to county in some areas.

Always remember: In any situation dealing with a contractor, check with the appropriate governing body—the state Contractor's Licensing Board, the state Department of Consumer Affairs, the local building department, or whoever has responsibility for overseeing contractor activity in your area—for full details on the laws and regulations that pertain to your project.

Basic Contractor Qualifications

In those states with licensing laws, some of the work involved in prequalifying a contractor as to ability and skill has been done for you. If your state or county requires a license and your contractor has one—which you'll want to verify—then he or she has at least met the minimum standards set by that region. If your area does not examine and regulate contractor qualifications, it's up to you to perform that task and assure yourself that the person you're hiring is qualified.

In states having licensing laws, typically the contractor must show and verify experience in the trades; pass a test that demonstrates knowledge of estimating, trade practices, contractor law, and basic business practices; and meet minimum standards of financial responsibility. In states that have registration laws, a person must at least show financial responsibility before being allowed to register as a contractor.

The minimum standards set by states and counties offer some valuable criteria to help you evaluate a contractor. You can use these state standards as guidelines in your own search for a qualified contractor.

Experience

One of the most important things to look for in dealing with a contractor is experience in the building trades—specifically in the area in which you need someone to perform. This doesn't mean that you should exclude a contractor who is newly *in business*—most contractors work in the trades for years before starting their own contracting business. But you do want to exclude those people who are newly in the construction field itself.

This may seem like harsh advice—after all, everyone has to start somewhere—but you're far better off finding someone who's already experienced in construction than allowing your job to be used as a training ground. In addition to having more practical knowledge as a result of years of hands-on experience in the field, established contractors also have a large investment in time, tools, and reputation to be concerned about. They have a business to operate and maintain, and your job—as well as your satisfaction with your job—is important to them.

One contractor signed a contract to build a new home for a young couple but had never built a thing in his life (some states do not require that a contractor have any building experience). He had, in fact, been a mobile home salesman until just a few weeks before. His crew consisted of his two sons, both still in high school.

The young couple discovered this contractor's lack of experience just as work was about to begin, but by then it was too late. The contract was legal and binding, and they were stuck. The house was completed—quite poorly—and he was nowhere to be found whenever they tried to contact him for repairs. A few simple questions early on would have saved this couple a good deal of grief, but they were taken in by a cheap price and a smooth line.

As a general rule of thumb, four years of experience in the construction trades should be considered a minimum. Using state board requirements as a guide, acceptable forms of experience can be strictly in the form of on-site work or a combination of work experience, apprenticeship classes, or even trade school or college-level classes in engineering, architecture, or construction management. Schooling alone, however, should not be considered adequate experience.

If you live in an area without a licensing law that checks and verifies experience, the only way to ascertain what experience a prospective contractor has is to simply ask him or her. If the contractor has been in business for a while, ask for references from past jobs. Call a couple of former clients to verify that they were satisfied with the job. Suppliers can also be contacted.

If the contractor is new in business but learned his craft while working for another contractor, ask for the name of the former employer. Call for a reference: Inquire as to what tasks the person performed, how long he or she worked there, and the quality of his or her work. It's doubtful you'll want to go so far as to check on college or apprenticeship school experience, but if the prospective contractor stresses an educational background as part of his or her experience, you might ask for the name of the school and the kinds of classes involved.

Skills and Knowledge

The next determining factor of a contractor's qualifications is his or her skill and knowledge in the trades. At first glance, this may seem to be much the same as experience, but there are some important differences.

Contractors who wish to become licensed in a state with rigid requirements first need to show—and prove—their experience in the field in order to qualify to take the licensing test. The test itself examines practical knowledge of common trade practices and materials, as well as knowledge of that state's laws as they pertain to contractors and their clients.

Years of experience do not always equal a high degree of either skill or knowledge—particularly in the field of work you might need to have done. For example, you may know of a restaurant that's been in business for years, but the cook still can't prepare a steak dinner the way you like it. He's fine at scrambling eggs and it's not a bad place to go if you want a quick breakfast, but it's not where you'd head for a special birthday or anniversary dinner.

The same holds true for contractors. Some have been in business for years specializing in fence repairs or building simple decks and porches—they're experienced contractors, but they may not be the ones you'd want to hire to build a new home or an addition to your house. Just as you wouldn't hire a plumbing contractor to do your electrical wiring, the general contractor you choose for a particular project should have skills and knowledge in those areas of the building trades that specifically affect your project.

Knowledge of the newest technology, materials, and trends in the field also plays a big part in a contractor's qualifications. If you're interested in remodeling your kitchen in the latest high-tech, European style, your contractor should have some knowledge of what that style constitutes, what products are available (or at least where to find out), and how to put it all together or find the people who can.

Financial Responsibility

The third important qualification a contractor should have is financial responsibility. This is a major concern even in states and counties with the most marginal and lenient of contractor regulations. The state may leave it up to you to determine if the person can actually nail two boards together, but most areas will at least try to ensure that you're protected financially—however marginally—in the event that something goes wrong with the job.

Most states will require that a person acting as a contractor carry both insurance and a bond; possession of these items is something you *definitely* should verify. Even if the state you live in does not specifically require that the contractor be licensed or bonded, you as an individual should insist on it. Dealing with an uninsured contractor is simply asking for trouble.

Insurance and bonding requirements vary from state to state and also change periodically. One of the first things you'll want to do—in fact, it should be done before you even call a contractor out to bid—is to contact the state or county agency in your area having jurisdiction over contractors and find out exactly what the insurance and bonding regulations and amounts are. Then you'll be prepared with the knowledge you need to ask the right questions when the contractors send in their bids (see Chapter 5).

Soliciting Bids

Ideally, you should get three competitive bids on a project. Comparing three bids will give you an accurate idea of costs, it will expose you to three different opinions and three different sets of ideas, and it will give you a chance to compare the contractors themselves to see who you would like to work with.

If you've worked with a particular contractor before and are confident of that person's abilities and prices, getting only one bid is fine. Otherwise, the problem with only one bid is that you just have one figure for the work, with no idea whether it's high or not.

Two bids, while obviously better, will give you only a high and low price; unless they're very close to each other, you'll still have no idea which is more accurate. Many people, despite advice to the contrary, will find meeting with two contractors (and following up) so time-consuming that they never contact a third bidder. In areas where building is booming, you may be able to find only two contractors with enough time to come out and bid the work, so you may have to be content with them.

On the other hand, getting four or more bids will often just confuse you, will take up an incredible amount of your time, and really won't benefit you in any appreciable way.

Deciding Who to Call

The first step is to make a list of the contractors you would like to have bid the work. Here are some suggestions for coming up with a list.

1. Word-of-Mouth Referral.   The first and best way to get recommendations is to talk with someone you know and trust who has used a contractor in the past with good results. Ask your relatives and friends, ask your co-workers and neighbors. See page 48 for some suggestions on what information you should try to obtain about contractors they recommend.

If you don't know anyone who's used a contractor recently, your next option is to ask strangers. If you see some work going on at a house you pass each day, stop and introduce yourself to the owners. Explain that you're thinking of having some work done on your house and ask them if they're satisfied with their contractor. Most people love to talk about their own construction projects, and if they like or dislike their contractor, you'll probably get an earful.

2. Ask Other Contractors.   If you have used a contractor in the past who is not bidding your present job—because he or she is too busy, is retired, or doesn't do this type of work—ask for recommendations. Even if it's just a plumber you once used to fix a toilet or an electrician who rewired your garage, these people are out on jobs and in contact with contractors all the time. They know who's honest and competent and who isn't.

3. Ask Material Suppliers. Another good source of names is the people who supply materials to contractors. If you're looking for a general contractor, stop by a couple of lumberyards and ask for their recommendations. Outline the kind of work you want to have done—different contractors do different things. They're sure to have names of several people they can recommend.

If you need a specialty contractor—a plumber, for example—look in the Yellow Pages for listings of companies that supply plumbing parts, especially at the wholesale level. Once again, call, tell them the kind of work you need done, and ask if they have any suggestions.

4. Ask Your Realtor. If you've just bought a building lot or a new house from a local realtor, ask him or her for recommendations. Realtors deal with contractors quite a bit, or at least know of others who deal with them; they should be able to give you a name or two.

5. Ask Your Insurance Agent. Insurance agents are often a good source of names since they work with a number of different contractors on insurance-related jobs. In addition, they often have a number of contractors insured through their office whom they know personally.

6. Check the Local Builders Associations. Most areas will have one or two trade associations that contractors can join for an annual fee. The association works to set contractor standards and polices its own industry, both locally and statewide. While these associations will not actually recommend a contractor, they do maintain lists of members. You can obtain this list at no charge.

Using a member of a builders association will not necessarily guarantee you a good contractor, but it will at least guide you to builders who are conscientious enough to belong to an impartial group that works for honesty and fair practices. You can find trade associations in the phone book or contact the Chamber of Commerce or the local building department.

7. Contact the Local Builder's Exchange. Many cities and towns maintain a builder's exchange office to which builders belong for an annual fee. The builder's exchange works like an intermediary that puts contractors and clients together. You bring in the plans and specifications for your work and they make them available to their contractors. Contractors come in at random (sometimes daily, sometimes almost never) and will get in touch with anyone having a set of plans that interests them. Place a time limit on how long the project will remain open for bid and request that interested contractors respond by that date.

The typical builder's exchange is usually heavy with large commercial and public works projects and new custom homes. Smaller remodeling projects are fairly rare.

8. Call the Local Building Department.   Depending on the functions of the building department in your area, they may be able to recommend someone who will do the type of work you need. At the least, they can usually provide you with a list of licensed and bonded contractors if that's required in your area.

9. Check with Other State and Local Agencies.   Local offices of the Contractor's Licensing Board, the Better Business Bureau, and the Chamber of Commerce will usually have lists of contractors available. Otherwise, they can at least tell you whether there are outstanding complaints filed against a contractor you're thinking of using. In the case of the Contractor's Licensing Board, they can also tell you if the contractor is currently licensed, bonded, and insured.

Making the Initial Contact

List in hand, it's time to start calling the contractors. Introduce yourself and tell them that you are soliciting bids for an upcoming construction project. Briefly describe the project, then ask them:

1. Do they do that kind of work? There's no point in wasting time with a contractor who doesn't do work in the field in which you're interested. For example, many contractors specialize in kitchen remodeling or room additions, while others may limit themselves to new houses and might not be interested in bidding a remodeling.

2. When would their current schedule permit them to start, if hired? Again, there is no sense in wasting your time or theirs if they're booked up for the next three months and you'd like to start in a couple of weeks. This question is especially important in areas with a strong local building economy—it will usually take longer to get a contractor.

3. Is there a charge for their initial consultation and estimate? Most contractors offer free estimates, although, in some of the busier metropolitan areas, there may be a small bidding charge. You'll want to clarify this up front.

Some contractors will charge a design fee if you have them work up plans and drawings for you and then don't use their services. This is a common and fair practice, since many homeowners request a full set of plans that represents hours of work for the builder and then walk off with the plans and give them to someone else or do the work themselves. If there is a design fee, clarify it with the contractor right away and find out if it is effective immediately or only if plans are prepared after the initial consultation.

4. Are they interested in bidding the work? Occasionally, for whatever reason, a contractor may be reluctant to bid a certain job. It may be too small, too large, too far out of town, or whatever, so it's best to extend the invitation to bid in this manner. It's obviously preferable to have a contractor who is interested in doing the job in the first place.

When you have gotten a positive response to all four questions, set up a meeting time that is convenient for you both. Contractors are used to bidding at odd hours—early mornings, evenings, weekends—in order to accommodate the schedules of the potential clients who contact them. If possible, arrange a time when both you and your spouse are available.

The Initial Consultation

When you hire a contractor, you are hiring much more than a body with a hammer attached to it. The signing of the contract will begin, for better or worse, a relationship with a person who will be a very big part of your life for the next several weeks or months. You'll see the contractor at work in your dining room while you're having your morning coffee. He or she will be in and out of your attic and your closets, eating lunch in your garage, using your bathroom and your telephone, hearing you fight with your kids, seeing what books you read and how much television you watch—in sum, learning about your life and becoming a temporary member of the family.

This is an unavoidable fact when you invite a stranger into your home for a long period of time. Contractors are used to the phenomenon, and, in fact, it's one of the attractions of the profession for many of them. With this relationship in mind, it's important to set a tone for the first meeting that puts you both at ease. You want to get to know this person a little, to get a feel for how well you can work with him or her. Some contractors, just like people in any trade or profession, may immediately cause unpleasant feelings, and you wouldn't want to hire such a person no matter how highly recommended.

When the contractor arrives at your home, be prepared. Have the plans ready, along with your notes, your bid sheets, and your list of questions. Set a casual, friendly atmosphere for the meeting, not a suspicious one. Remember that no matter how often a contractor goes out on a bid, he or she is still a little apprehensive upon entering a stranger's home for the first time. The contractor is trying to make a good first impression and you're hoping to find a person you can trust—an initial strain for both of you that may take a little time to get over.

Many homeowners—entering into a world of unfamiliar terms, myriad choices, and sometimes overwhelming expenses—are often intimidated by contractors. Most are afraid to appear "stupid" by asking too many questions. Even those who are willing to ask questions are usually so unsure of exactly what they want to have done that they really don't know which questions to ask anyway.

If the contractor is there to bid a remodeling or repair project, give a brief tour of those parts of the house that relate to the work and explain your ideas as you go. Don't be offended if he or she begins looking into closets and under sinks—it's all part of the process of evaluating your project.

If he or she is meeting with you to bid the construction of a new house, you'll basically just be going over the plans and discussing your ideas. Therefore, you might meet at a restaurant or at the building site if you'd prefer.

After the house tour, sit down and go over your notes and your bid sheets in detail. Take as much time as you need—or as much time as the contractor has available at that time—and explain what you want to do. Mention when you'd like to start and if you have a particular deadline you need to meet—guests coming from out of town, for example.

Next, explain what work you would like to do yourself—if any. Some contractors are less willing to work with homeowners in this area than others, and it's important to get a feeling for how receptive the contractor is to this. If it's important to you to do some of the work yourself and the contractor seems reluctant to cooperate, it may be necessary to find someone else to work with.

At some point during the meeting, it will be necessary to discuss your budget. It's much better to be honest about what you can afford to spend than to waste everyone's time waiting for a bid that's way out of your price range.

One couple invited three contractors to their house, carefully discussed all the many details of the room addition they wanted, and even provided the bidders with a specification and materials list. Only the last of the contractors inquired about budget, and the homeowners were shocked to hear that the $20,000 they had available for the project would fall far short of being enough to do all they wanted. This contractor suggested changes, but it wasn't until they saw the bids from the first two contractors that they were convinced the third one was right to be concerned about their budget. They had to revise their ideas, request new bids from everyone, and finally gave the job to the third contractor. If they had been willing to discuss budget as part of their initial consultations, the entire bidding process would have been accomplished much more easily.

If you are unprepared to discuss your budget, you can ask the contractor for some idea about how much the project will cost. Most contractors are reluctant to make off-the-cuff estimates—if they quote a figure that turns out to be too low, the homeowners think they're being cheated; if the number is too high, they may scare the homeowners off and not even be asked to bid the job. But they may give you a range of prices so you can see if it's within your reach. That way, neither party has to commit to a definite figure.

Some Questions to Ask

While the contractor is there, be sure to ask for his license number and for details about bonding and insurance. If the contractor is unwilling or unable to provide you with this information, thank him or her politely, adding that you only want to deal with contractors who are legally licensed. *Do not*—under any circumstances and no matter how attractive the price—deal with an unlicensed or uninsured contractor. If you do, you will be taking a *tremendous* risk. With the number of competent, licensed contractors available, it's foolish to deal with someone who isn't even honest and competent enough to comply with the law.

Next, ask if the contractor would be willing to provide you with the names of some former clients whom you can contact. You will most likely be given only the names of *satisfied* customers, but even talking with them will give you insight into how the contractor operates and how good his or her finished product is. If the contractor does not have this information at the meeting, ask that it be provided with the bid.

It is also a wise precaution to ask for a bank reference and for a few businesses where he has accounts. Most homeowners are very reluctant to ask this, but it's in your best interest not only to request the information but actually to verify that the contractor is solvent and financially responsible. You are, after all, going to be entrusting this person with a great deal of your money, and, in many cases, you remain legally liable for paying any bills that the contractor neglects to pay, even if you've already paid him or her in good faith (see Chapter 5).

It's completely acceptable to inform the contractor that other people will be submitting bids on your project—they're accustomed to competing for work. All they ask is that you provide them with the necessary details and keep them informed of any major changes that may arise in the project so that they can adjust their bid accordingly. Don't ever schedule two contractors to meet you at the same time. Some people may tell you that it's a clever ploy to get the builders to compete against each other, but all it's going to do is anger the contractors and prevent you from discussing your project adequately with either of them.

Talk with Former Clients

While you're waiting for the bids to come in, call one or two of the client references the contractor gave you. After introducing yourself, explain why you're calling—that you are about to start a construction project and the contractor bidding the job gave their names as a reference. Naturally, many people may be suspicious at first and may not be readily forthcoming until you've put them at ease.

If you feel comfortable with the person after talking for a while, you might ask if you could come by and actually see the contractor's work. This will give you a first-hand look at what the contractor did and how well it was done.

In order to be sure that you get as much information as you can without imposing on the person you're talking to, write down a list of questions before calling.

Here is a list of suggested questions to ask (you will probably want to add specific questions relating to your job):

- Were you happy with the overall job that the contractor did for you?
- Did the job start on time? Was it completed on time?
- Was the job done for the estimated price, or were there a number of additions and change orders that ran up the final figure? (The answer to this can be misleading; the additions and change orders could all have been at the request of the homeowner, not the contractor.)
- Did the contractor live up to the contract and to any promises or assurances made prior to the job's starting?
- How was the contractor to work with? Was he or she flexible with changes and patient with questions?
- How competent was his or her crew? Were they easy to get along with? Were they courteous and careful while on the job?
- How skilled were the subcontractors that were hired?
- How well coordinated was the scheduling? Did materials and subcontractors seem to be on the job at the right times, without a lot of overlaps or unreasonable delays?
- Did the contractor stay in communication with you? Did he or she return phone calls and show up for meetings at the appointed time?
- Would you hire this contractor again?

Compare the Bids

When you have all your bids in hand, it's time to take a long, careful look at them. There are a number of things that you'll want to look for and compare, and it's important that you be objective. Even if you think you might have already decided which contractor you'd like to hire, it's still extremely important to review all the bids to make sure that nothing was overlooked, and that the price is competitive.

Compare the Presentations

This is a fairly minor thing, but it's worth taking a moment to look at and compare the overall presentation of the bids. Is one quickly handwritten on scrap paper? Is one neatly typed on letterhead? Did any of the bidders include extra materials, such as catalogs, references, copies of articles relating to the work you want done, or anything that shows a little extra effort? As you compare the bids, look to see if one is more organized and clearly spelled out than another.

While these fairly minor differences may not amount to anything in the long run, they do contribute to the first impression the bidder makes on you, and they may have some bearing on how seriously he or she takes the profession and how much pride he or she takes in the work.

One woman tells of the contractor that she called for a bid on a complete kitchen remodeling. He asked her to describe the details of what she wanted over the phone so that he could give her a bid without having to come out and see the job.

Another couple described the contractor who was called out to bid on a small room addition and after five minutes, with virtually no discussion of details, told the couple he could do the work for $10,000. As the contractor left, the homeowners requested that he submit a written estimate for the job, with a breakdown of what he was going to do. A week later, they received a handwritten proposal in the mail that read: "One room addition, 10 × 16 feet. $10,000."

Compare the Prices

No matter what other considerations people might have about their job, price is usually the most important, so this is as good a place as any to start the comparison.

Look at the bottom line price of each bid. Because all the contractors are from the same area, using basically the same rate scales and the same materials, there shouldn't be a huge variation in prices. All should be relatively close to one another, within about 10 percent one way or the other.

If one of the bids is considerably lower, it's likely that the bidder overlooked something. For example, if one bid came in at $20,000 and the second came in at $21,200, it's highly unlikely that you can expect the same quality of labor and materials from a third bidder at $14,500. That bidder completely forgot something, unintentionally bid the wrong materials—aluminum windows instead of wood perhaps—or is being underhanded in trying to lure your business with a low bid that he or she will jack up later with lots of change orders.

If one of the bids is considerably higher than the others—say $28,000 against bids of $20,000 and $21,200—then that bidder accidentally misbid something; has thrown in something you don't want or need, often intentionally; or has purposely bid high.

The practice of intentionally bidding high is actually fairly common if the contractor doesn't really want a job. It may be a dirty or disagreeable project; or perhaps the contractor foresees a lot of potential problems; or the contractor may be so busy he or she doesn't really need another job right now; or he or she may sense that you will be difficult to work with. Whatever the reason, the contractor may intentionally overbid the work; the reasoning is that if the job goes elsewhere, the contractor doesn't care, and if he or she does get it, it will be worth doing because of the money.

Compare What's Included

Now look at what each contractor has included. The bid should be well detailed and should clearly spell out as many details as possible about the work to be done and the materials to be used. Here are some things to look for:

• Specified sizes are correct—for everything from the overall size of a room addition to the sizes of individual windows and doors.

• Quantities should be correct. If the plans call for six doors, then six doors should be listed.

• Brands and model numbers should agree with any that you specified in your initial bid sheet. If you wanted Acme oven number A-1, be sure the contractor didn't bid Acme B-2 in order to drop the bid amount a little. Remember: Once it's specified and agreed to, you're stuck with it—if you want the A-1 instead of the B-2, you will probably have to pay the difference later on. A low bid really isn't worth much to you if the materials you really want cost extra.

• Material grades are what you specified. If you wanted oak trim, be sure it says oak trim and not just trim. If grades were *not* originally specified, contractors may include the material they normally use or that they think would be most appropriate in your situation. Be sure to compare what one contractor has specified against what the others have specified; this can make a big difference in prices between bids.

Compare the Bid Allowances

One thing that confuses many homeowners—and where some less-than-honest contractors take financial advantage—is the subject of bid allowances. These are very common where the exact final specifications have not been worked out at the time the bid is prepared, especially for decorative items such as carpet, wallpaper, tile, and light fixtures.

A bid allowance means that the contractor has placed a certain dollar figure in the bid to cover the cost of a certain item. When it comes time to select that item, you will be allowed the specified amount of money toward its purchase—if you select a higher-priced item, then you will have to make up the difference; if you happen to select one that's cheaper, then you're entitled to a refund of the difference between the actual price and the allowed price.

For example, your bid may read "carpet allowance, 80 square yards, installed—$20/square yard." When you go looking for carpeting, that's the amount you have to spend. If you find a carpet you want that's $22 a yard, you'll be responsible for the

$160 difference ($2 per yard × 80 yards). On the other hand, if you like a carpet that's $18, you should receive a $160 credit on your final bill.

Another example might be "lighting allowance, 10 fixtures, installed—$500." In this case, your *total* allowance for those ten fixtures is $500, and you can break it up any way you like. If you select a chandelier for the dining room that's $400, you're left with $100 for the nine other fixtures—unless, of course, you are willing to pay the additional amount for selecting more expensive lights.

While bid allowances are a very common and necessary part of the bidding process—since you certainly can't expect to have selected all your lights, floor coverings, appliances, and a hundred other items before your new home or remodeling project has even begun—the process does allow the shady contractor a few opportunities for deceit. Here are some suggestions for your protection:

1. Get the Allowance Amount in Writing.   If you are being allowed a specific amount for a certain item, be sure both the item and the amount are clearly stipulated in the bid. For example, the bid should read something like:

> Appliance allowance, installed, as follows:
> | Range | $800 |
> |---|---|
> | Built-in dishwasher | $300 |
> | Trash compactor | $350 |

Failure to get the amount clearly in writing can easily lead to problems later on—both honest and dishonest ones. For example, the contractor may have legitimately put $300 in his bid for a dishwasher simply because that's what he's paid for the last ten he's installed, but the one you want is $400. If it isn't specified in writing, which one of you pays the additional $100? This is obviously a gray area, and arguments frequently result.

Not getting it in writing can be worse if the contractor is unscrupulous. No matter what was actually figured into the bid, if it's not in writing, the contractor is going to tell you that it was equal to or less than the cost of the item you selected. The contractor may have calculated $500 in the bid from the start for that $400 dishwasher, but if you haven't specified an amount, he or she can either keep the $100 difference or tell you that only $300 was allowed and ask you for an additional $100, skimming a total of $200 off the transaction.

2. Make Certain the Quantity Is Specified.   If the contractor is allowing you so much a square yard for carpeting, the bid needs to specify how many square yards he or she intends to supply. The bid should read: "Carpet allowance, installed, with pad, 90 square yards—$20 per square yard."

Like an unspecified dollar amount, an unspecified quantity can lead to misunderstandings—or worse. The contractor can say that he or she allowed for installing only eighty yards of carpet—perhaps claiming: "I didn't allow for the closets because nobody wants carpeted closets anymore," or "I didn't figure in carpet for the guest bedroom; I thought you said that you wanted to do that room later." Either way, you're left holding the bill for the extra ten yards of carpet or forced into a legal battle to get your money's worth.

3. Be Certain That the Specified Prices and Quantities Are Reasonable. One trick that an unscrupulous contractor will use is deliberately to estimate low on either the price allowance, the quantity allowance, or both. This allows the contractor to present you with a lower bid for the work than the competitors, while still not risking any loss on the contractor's part. If you agree to an estimate that contains low allowances, you're stuck with it. You either use cheap materials or pay what could amount to thousands of extra dollars in upgrades.

This is where a little common sense and some simple shopping on your part will come in very handy. Visit a few stores, check some mail-order catalogs, and read the ads in your local newspaper—you'll quickly get a pretty good idea of what various things cost. If you found some dishwashers you liked that were all in the $350 range, be sure the bid doesn't say $200. If the carpets you're attracted to are all priced around $20 a square yard, see that the bid reflects that. You can always switch to a less expensive appliance or carpet later if the bid turns out to be higher than your budget will allow, but in the meantime you're getting a bid that's honest and realistic.

It is possible, of course, that your tastes run to materials that are more expensive than even the honest contractor normally allows. For example, $20 to $25 may be all the contractor ever allows for carpeting, since that's the price range that satisfies most of his clients. If, upon shopping around, you find the carpeting you want averages $10 or $15 a yard more than that, be sure to tell all of the bidding contractors to increase their allowances accordingly. That way, the bids remain competitive and you're not shocked at the end of the job when you're expected to pay several thousands of dollars in legitimate allowance overcharges.

As far as quantities are concerned, do some quick calculations on your own to see if the contractor is allowing for enough material. If your room addition is 20 × 30 feet and it's to be completely carpeted, some simple arithmetic will tell you that the addition is 600 square feet, and that you will need approximately 67 square yards of carpet (there are 9 square feet in a square yard, so just divide 600 by 9). There may be other things for which the contractor is adding or subtracting carpet—stairs, odd angles, patching, etc.—but if only 50 square yards have been allowed for, have the contractor explain how he or she arrived at the figure.

Similarly, look at the plans and count the number of light fixtures being used; quickly calculate how many square feet of tile will be in the entryway, or the size of the walls that are to be wallpapered, or how many doors there are. If the contractor's figures don't look right to you, ask for clarification.

4. If the Bid Allowance Includes Installation, Make Sure It's Specified. Be sure that your bid specifically states whether the allowance given includes installation. This may seem obvious, but this, too, is a real gray area that can cause problems, so don't take anything for granted.

When carpeting is advertised at so much per square yard, remember that that price is almost always for the *carpet only*. Installation can often run an additional $3 to $4 a yard or more, which can increase the cost quickly.

5. Specify Who Will Select and Pick Up the Materials in Question. It's a good idea to have the contract specify who will select the materials and who will actually pick them up and get them to the job site. Again, this will eliminate any confusion and potential arguments later on.

In most cases, you will be selecting the materials—after all, that's the primary reason that an allowance was placed in the estimate in the first place. In the case of carpeting, you would go to the carpet store and select the carpet you want, and the contractor would arrange for its delivery and installation. Light fixtures, bathroom accessories, and other smaller items may be picked up by the contractor or by you, depending on the agreement.

The contractor may also request that you shop at certain stores. This is not unusual and should not become a cause for suspicion. Contractors become familiar with the quality of materials and installation they can expect from certain shops, and prefer to continue working with an outlet they know and trust. They also know and can depend on the exact percentage of contractor discount they're allowed, and they may have an account already set up there. If you would like to shop elsewhere, inform your contractor so that any necessary arrangements can be made.

You will probably find that most of the shops you deal with—whether or not the contractor has sent you there—will quote you only the full retail price of an item, not the contractor's price. Some outlets deal only with contractors, meaning that the general public cannot shop there except by going through a contractor. Other places have both retail and contractor prices, and quoting you only the retail price is common practice.

Select the Contractor

Now comes the moment of truth. You've talked with the contractors, you've compared their bids, and now it's time to make the choice. For some, the choice was made within five minutes of the contractor's arrival. For others, there may still be some indecision even after several meetings and a careful review of the bids.

The decision as to which contractor to choose is a result of subjective and objective opinions. Here are some things to ask yourself about each contractor:

• Do you think you can work with the contractor? This is a crucial question, since this person will become a part of your life for a while. Ideally, you would like to find a person who is friendly, knowledgeable, competent, and flexible. All these traits will combine to make the project easier and more enjoyable for you.

• Is the price fair? Do not be tempted to immediately jump at the low bid or reject the high bid. There are many things to consider when looking at a job; price is only one of them.

• How did former clients feel about the contractor? Was the job done on time, with good results, and a minimum of delays and extra costs? Was the contractor amenable to making changes when necessary?

• Did you receive help with your design questions? A surprising number of contractors are content to wander through the job, listen to your comments, and bid whatever you tell them without offering any input of their own. More experienced contractors will often take the time to talk with you and offer suggestions of alternate ways to do things—either to obtain a better finished product or to lower the cost. This goes back to finding a contractor you can work with. A person who's willing to talk and answer questions can make your project much easier for you.

• Did the contractor seem to be honest with you? Were you provided with references, license numbers, and other information when asked, or did the contractor seem hesitant to cooperate? Were there schemes or suggestions or evasions that made you doubt his or her honesty?

When you think you have made your choice, take a moment to verify his license number, his insurance, and his bond. Call the number on his business card and be certain it's a legitimate number. Remember that many contractors work out of their homes and don't maintain formal office space, so the address and phone number given on the business card may well be his or her home.

When your choice has been made, notify the chosen contractor and set up an appointment to review the project and sign the contract. As a courtesy, especially if you think you may have need for their services later, it's always a nice idea to call or write to the other bidders, thanking them for their time and their input.

Insurance Repair Work

The bidding process will vary from what's been described so far if you are having insurance-related repairs done. Contractors bidding insurance jobs have to deal not only with you but with the insurance company, too. Therefore, the procedures are often different.

If you have had a fire in your home or sustained some other loss that is covered by your insurance, the first visit will be from your insurance adjuster, not a contractor. The adjuster will go over the damage with you, determine what is and isn't covered, go over your deductibles and depreciation, make notes and measurements on the current condition of the home, and discuss your options with you.

After the adjuster is finished, you will need to contact a contractor. Your adjuster may be able to suggest a few local contractors who do the type of work involved in your loss, or you can contact any contractor using the guidelines suggested on pages 42–44.

The contractor will visit your home to view the damage and estimate the repairs. He or she will need either to see the adjuster's scope of work—the forms the adjuster filled out describing what needs to be done—or to meet with the adjuster on the job. Either way, the contractor needs to have a clear idea of what portions of the loss are covered and what the adjuster will be willing to pay for.

Bidding insurance work is tricky for the contractor. The adjuster may say that the damage was contained in a certain room or area of the house, but in order to make a proper repair, the contractor needs to extend his work into an adjacent area. The adjuster decides how much will be allowed, as well as the prices and amount of work to be done. You should participate in any meetings between the contractor and the adjuster so that you have a clear idea of exactly what's covered and what isn't.

The contractor's bid will include a number of allowances, and you should follow the same precautions outlined on pages 50–53. Remember: All the choices are yours—carpet, lights, appliances, etc.—just as if you had hired the contractor for a remodeling project. The only difference is that the adjuster will set the bid allowances, based on what was previously existing at the time of the loss. You will be responsible for paying for any upgrades.

Remember that the insurance company cannot *force* you to deal with a particular contractor. They can suggest contractors they have worked with in the past whom they know to be competent in the type of repairs you need to have done, or you can call in a contractor whom you know and trust. The work to be done will be defined by the insurance adjuster, but it's still important that you receive competitive bids.

Once the contractor has been selected and the price has been agreed upon, you will need to get everything in writing. Remember: The contract will be between *you and the contractor,* not between the contractor and the insurance company. The insurance company will pay for the loss, less your deductible amount, but in virtually all situations you will be ultimately responsible for paying the contractor. Follow all the suggestions given in Chapters 4 and 5 for preparing a contract and protecting yourself against liens.

Scams and Swindles

It is an unfortunate fact of life that many people have had problems with contractors over the years. There are mild headaches, major problems, and out-and-out swindles. State regulations that mandate licensing help considerably, but there will almost certainly always be homeowners who fall victim to building contractors.

By far, most contractors are both honest and capable. It should be noted that most construction jobs run their course with no more than minor problems, and some of these must be laid at the feet of the homeowner. But setting the majority of honest, capable builders aside, there still remains a fairly sizable group of problem contractors who manage to keep the building trades at or near the top in the number of consumer complaints filed each year.

This problem group can be broken down into three basic categories. You can prevent yourself from falling victim to any of these people with simple common sense and some equally simple precautions.

Incompetent Contractors

Incompetent contractors bid the job wrong, do the job wrong, mismanage their finances, and hire equally incompetent employees and subcontractors. The homeowner is all too often left with an unfinished job or a bundle of costly repairs.

This is the area most affected by state regulations. Requiring and verifying experience in the building trades is the best way to weed these people out. If the state won't do it, then the burden falls on the individual consumer. When dealing with any contractor, ask to see some completed jobs or at the very least speak with previous clients. Also ask for some financial references—lumberyards, wholesale houses, subcontractors—and call a few.

Somewhat Dishonest Contractors

The second group consists of the marginally to moderately dishonest. Here are some examples:

• They will deliberately underbid a job in order to get the contract, then manipulate the homeowner with change orders until the final bill is higher than the highest of their competitors' original estimates.

• They will substitute inferior-quality materials for the ones that were originally specified and for which you're being charged. In extreme cases, they may use old materials that they removed from someone else's job.

• They will do only part of the work agreed to while charging you for all of it—applying one coat of paint when you paid for two is a common example.

• They will cheat on time-and-materials jobs—that is, when you are paying for the actual hours worked and the actual materials used as the job progresses rather than having established a firm price before the job starts. They will pad the hours, pad the bills for materials, or both. You may be charged for the hours they're at lunch, or even for hours they're on someone else's job. You may be shown a legitimate bill for lumber from the lumberyard, but some of the materials you're being billed for were returned, went to another job, or even went to their own house.

Truly Dishonest Contractors

The last group includes the out-and-out con artists, the ones with the well-thought-out schemes, the ones that do the most damage to consumer pocketbooks and contractor reputations.

Over the years, there have been many contractor con games. Some you may have heard of, some may surprise you. Here are nine of the more innovative ones to guard against.

1. The "Model Home."   In this scam, typically for house siding, insulation, solar products, or a combination of all three, the fast-talking salesperson offers you siding at a cut-rate price, say $2,000, if you'll agree to allow your house to be used as a model for the neighborhood. For every house that he does in the area as a result of the people having seen how great your house looks, you'll get a rebate of $200. Help him sell just ten jobs and your siding will end up costing you nothing. Sign today and he'll even advance you the first rebate amount—he's that confident you'll be selling more siding jobs for him.

So you sign for an $1,800 siding job, with a big down payment. He does indeed side your house, but with a vastly inferior product than what he had shown you during his sales visit. And the "model home" rebates he promised? There are none, since every other house in your neighborhood that he's sided is also supposedly a model house.

2. Leftover Materials.   A man knocks on your door one day and tells you that he's a contractor who specializes in sealing asphalt driveways. He presents a business card, and when you glance outside you see a reassuringly large truck with his company name on the side. He offers you a great deal: He has just finished a sealing job up the road, but he accidentally brought out too much material, which will spoil if left on the truck for the drive back to his shop. The two of you can really help each other out: Rather than waste it, he'd like to go ahead and seal your driveway for you, and he can do it for a fraction of the normal cost, say $300 instead of $700.

It sounds like a great deal to you, so you agree. His men jump off the truck, they do the job, your driveway looks shiny and black, and the man comes back and asks for $500. It was supposed to be $300, you protest. He claims you misunderstood him—there's nothing in writing, of course—so he offers you another favor and agrees to split the difference. You give him $400 and then find out the next day that all he put on your driveway was watered-down black paint, or perhaps used motor oil.

This scam has been around for so many years it's amazing that so many people still fall for it, but hundreds do every year.

3. The Cash Discount.   A contractor sells you a remodeling job, new siding, or new roofing for $5,000. She tells you that if you'd be wiling to pay for the job up front, in its entirety, she could pay cash for the materials and get a big discount. That discount would, of course, be passed along to you, so she could do your job for a mere $3,500 instead of the originally quoted $5,000. Another great deal, so you reach for your checkbook. Any big surprise that you never see the contractor or your money again, and that the address and phone number on the business card don't exist?

4. Bait and Switch.   You hear about these scams all the time, and while they're certainly not limited to contracting businesses, they are a favorite of dishonest contractors. You see an advertisement for a summer sun porch with a nice roof and screened panels for the unheard-of price of $200. When you contact the company, you're told that the illustrated porch is the "deluxe" version, and a smooth talker quickly convinces you that you'd be much happier with that model—it's vastly superior to the basic model, and even though it's $1,200, she can tell you're a person of good taste who really appreciates quality.

She's sorry you misunderstood the ad, so she'll even do the deluxe porch for $1,000 if you don't tell any of your friends how you got her to work on your house "as a favor, without any profit." Order the deluxe model, and you'll get a cheap enclosure that looks something like the picture. Insist on the basic model advertised, and you'll get a few 2 $\times$ 4s strung together and nailed to the house, wrapped with a little bit of cheap window screen.

5. The Contest.   Lucky you, you've been selected in a random drawing and you've won $500 worth of blown-in insulation—materials only. Of course, the insulation can be installed only with the special blowing equipment that the contractor owns, but he can provide the labor to do the job at a low, low, contest-winners-only special price of just $400. You agree, figuring $400 is a bargain for an attic full of insulation.

When he gets to your house to do the job, he finds that, given the size of your attic, the $500 worth of free insulation will cover only half of it. If you want the other half done, it will be another $400 in labor—he'll still give you the discounted price, since he's already there—plus another $500 worth of additional insulation.

You either settle for half an attic or pay a total of $1,300 for the whole attic—more than you would have paid a reputable contractor to do the job in the first place.

6. Phony Inspectors. Yet another knock-on-the-door scam, this one involves a phony building, electrical, plumbing, or fire inspector, or a representative of the gas or electric company, complete with authentic-looking credentials. The inspector will tell you about some problem in the area that is associated with houses constructed around the time yours was built—leaking gas piping, shorts in the electrical system, or perhaps a bad chimney.

The "inspector" will insist, sometimes with an urgent sense of impending doom for you and the entire neighborhood, that your home be inspected for problems of a similar nature. The problem will of course be found, and then the "inspector" will insist on doing the repairs immediately because of the extreme danger. In a couple of hours, you'll be presented with a whopping bill for some minor repair or adjustment somewhere, or even for work on your home that actually left it in worse shape than before.

A common variation on this is the furnace inspector, with equally important-looking credentials, who will report furnace problems in the area and ask to check yours, which will need repair or, worse yet, complete replacement. In the complete replacement scam, your furnace is removed and replaced with one that's been "reconditioned"—the claim being that they can sell you this one cheaper than a new one. Your reconditioned furnace is actually the one they just pulled out of your neighbor's house, and your perfectly good original furnace will then be passed along to someone else in the area in a perpetual cycle of phony reconditioning.

7. Financing Schemes. After being sold a repair or remodeling job—perhaps one you didn't need in the first place—the contractor will lead the discussion around to finances. He'll mention—with an air of confidentiality and "I think I can trust you folks"—how he has helped people out in the past by falsifying estimates and bank documents in order to get a larger bank loan for the homeowner. For example, he might tell you that since the estimate is for $7,000, he could easily make it $10,000, leaving you with an extra $3,000 for paying bills, taking a vacation, or whatever.

At this point, there are two basic variations. In one, he has you sign a contract for the full $10,000, then claims at the end of the job that that's what you actually owe him. You got $7,000 worth of work done, if that, for $10,000, and you don't really have a leg to stand on. In the second variation, he insists, for tax purposes or whatever, that the entire loan be deposited to his account, and he will then turn around and write you a check for the extra $3,000, leaving the original amount of $7,000 in his account with which to do the job. He will then simply take off with either the $7,000 or the full $10,000, leaving you with nothing but the loan payments.

8. The Silent Partner.   This is another type of financing scheme, in which the contractor gets you to become a silent partner in the construction of your own house. There are some rather complicated papers in legal jargon to sign, but the upshot is that the contractor offers to pad the estimate for the bank and help you secure a larger construction loan than you actually need to build the house.

The contractor will then pad the payment request, pay you the difference between the real payment due and the padded payment received, and you will have the bank's money to use, at a lower interest rate than you could get with a straight loan, until the house is completed. The trouble is, you get only a fraction of the padded amount, and the contractor takes off with the last payment in its entirety before the house is done. You're left with an unfinished house that, even partially done, has already cost you much more than a fully finished house done by an honest contractor.

A variation on this is the padded insurance estimate. You hire the contractor to do some insurance-related repairs, and he or she offers to pad the estimates and the bills to the insurance company. Insurance company checks are two-party—in your name and the contractor's—so the contractor will suggest that you sign them over to be deposited in his or her account "to keep you in the clear in case the insurance company gets suspicious." You get a partial repair job, or none at all, and the contractor is gone with the money.

9. The Extras.   Here's another very common scheme with a hundred variations that draws people in by the thousands every year. A contractor comes out to your home and gives you a "firm" price on having something done—new siding, for example. He may even offer a contract for the agreed-upon amount, with the wording in the contract saying something along the lines of "provide and install new siding for the entire house." You have a price that's cheaper than his competitors and you have a contract, so you agree to have the work done.

You come home one day and the contractor says the work is done—the siding has been installed on the entire house. What hasn't been installed is any trim—there are jagged cuts around the windows, and gaps where the corners don't meet. You ask why there's no trim and are told that you didn't ask for any. The price and the contract do not specify anything about trim. Sorry about the misunderstanding, we'll be glad to install the trim for an extra $700. The price is now well over that of the contractor's competitors, you're stuck, and with the signed contract there's enough of a gray area that you'd spend a small fortune fighting in court—considerably more than the extra you have to pay.

In Chapters 4 and 5 you'll see how to protect yourself against occurrences like these, but it's worth touching on it here, too. Keep in mind that these schemes are not always that obvious when they're happening, and there are many reasons that thousands and thousands of people are swindled each year. These con artists are very smooth talkers with an answer for everything, making even the most unctuous politician pale by comparison.

Their targets are often the poor and the elderly, although they have scored victims in every age and income group. They prey on fear, money, anger, jealousy—anything they can read in you that they can exploit—and their pitch is not always an easy one to resist.

For one thing, everyone loves a bargain. The thought of buying someone else's leftover materials at a huge discount is enticing, as is the thought of having your home sided "for free" with the rebate checks from your neighbors' sales. And that big "cash discount" if you pay more money up front—you were going to pay the contractor anyway, so why not sooner? The old adage, which everyone has heard but often flies out the window when it's most needed, is that "If it sounds too good to be true, it usually is."

With the bank and insurance schemes, there's a combination of greed and something akin to revenge at work. The banks and the insurance companies have ripped you off for years, right? They're always there with their hands out, making a profit on your hard-earned money, so why not take this contractor's offer and turn the tables on them? It's been done before, and there's no real risk—why, you're only making use of your own money anyway.

Then there's the fear of authority. When that exterminator or that inspector knocks on your door with tales of voracious insects and exploding furnaces, not giving you any time to really think because of the implied urgency of it all, it's hard for some people to close the door. Try not to let yourself be fooled—business cards are cheap to print, and anyone can order official-looking shirts, signs, ID cards, and even badges by mail.

4

Understanding the Contract

When you undertake the purchase of a new car, you have the car itself as physical proof of what you are buying. Even at that, the dealer will give you a written list of what the car contains—the size of the engine, the type of transmission, even the kind of radio—as well as a number of papers clearly explaining all the financial terms of the deal. Finally, all of the terms of the warranty will be spelled out in black and white. The average car buyer, feeling proud of his or her savvy, would accept nothing less.

With the undertaking of a construction project, you are buying nothing more tangible than the blue lines of the plans and the professed expertise of the contractor. And yet a surprising number of homeowners are willing to accept only vague written details and a handful of verbal promises before committing enough money to buy two or three new cars.

With any construction project, large or small, it is imperative that you get the details of the agreement in writing, with as many specifics as possible. To do less is to open yourself up to problems of all kinds and is, in a word, just plain foolish.

The specific laws relating to contracts are different in each state, giving both you and the contractor a variety of legal rights and responsibilities. If you do not understand *any aspect* of the contract you are being asked to sign, and if the contractor cannot explain it to your satisfaction, it is best to consult with an attorney before proceeding further.

What Is a Contract?

In a broad sense, a construction contract can be virtually anything that you sign with the contractor. If the contractor presents you with a written proposal, your signature on it may constitute your legal agreement to have him or her do the work, as well as your acceptance of the price and specifications that the proposal contains.

Many contractors use simple, stationery store–type proposals and contract forms. These forms are in plain English and easy to understand, but they commonly lack details and information from a consumer's point of view. If the contractor presents you with a form contract that is vague or oversimplified, you can use it as a starting point, asking that it be amended as necessary to further spell out the details of your agreement.

A preferable alternative to the basic, prepackaged contracts are the more detailed, attorney-prepared contracts that other contractors use. These are typically much more detailed and are written to encompass the pertinent laws of a particular state. Once again, though, contracts such as these are not always the perfect solution. They tend to be written in a way that often favors the rights of the contractor and may need to be amended to offer adequate protection for the consumer.

This is not to say that your contract needs to be overly long. Some attorney-prepared contracts are so long and so mired in unintelligible legal phrases that they are virtually incomprehensible to the average person. They need not be this complicated, they just need to be complete enough to spell out your rights and obligations and the specific details of your agreement.

A Few Basic Rules

There are a few basic rules you should always follow with regard to contracts of all types. Most reflect simple common sense.

Don't Sign Anything Until You're Ready.   The first and most important rule with proposals, contracts, or any paperwork presented to you by the contractor is not to sign *anything* until you are fully satisfied with what it says and you are actually certain that you want to proceed with the work. Some contractors will try to get you to sign the estimate sheet or a proposal of some sort—"just to show you're interested" or "so that I have something to show my boss"—then they pressure you later by saying you already agreed to have the work done.

It bears repeating that *practically anything you sign could be taken as an agreement and contract,* especially if the fine print is ambiguous or misleading. On a typical construction job, the only things you should ever need to sign with a contractor are the actual contract—and then only when you're ready to commit to the project—and any change orders if and when necessary as the job progresses.

Never Sign a Blank Contract.   You should never, *under any circumstances,* sign a blank contract or any form that the contractor presents to you. Don't ever fall for an explanation such as "rather than delay the job, why don't you sign the contract now so I can get the crews started—we can fill in all the specifics later." In all fairness, most contractors don't work this way, but there are those exceptions to guard against.

On a similar note, don't ever sign a contract that has blank spaces of any kind—every blank line should be completed. If the blank space refers to something that doesn't apply to your job or your situation, it should either be deleted or marked "not applicable" (or N/A). Once again, don't ever let contractors tell you that they'll fill that space in later.

Be Sure the Contract Is Complete. When you are ready to sign the contract, be certain that it spells out your *complete* agreement with the contractor. All of the things you talked about should be listed as thoroughly and accurately as possible.

Never Sign Anything You Don't Understand. If you do not understand any portion of the contract, don't sign it. Ask the contractor for a complete explanation of those things that are vague, incomplete, or worded in a manner that makes you hesitant. Have the contract changed as necessary until you're comfortable with it.

If the contractor cannot explain any part of the contract to your satisfaction, delay signing it. Consult with the Contractor's Board, another contractor, or an attorney to clarify the problem areas.

Get an Exact, Signed Copy. Be sure you get an *exact* copy of the contract you sign. Most contractors use a two- or three-part carbonless contract form so they can give you a copy immediately. Your copy needs to be an exact duplicate of the contractor's copy, and it needs to be signed and dated by the contractor. If, after signing, there are changes to the contract, be certain the changes are done in ink on both your copy and the contractor's, and be certain that *both* of you initial the changes on *both* copies.

What Your Contract Should Contain

Along with these general rules, which apply to any contract, there are a number of specific items unique to a construction contract of which you should be aware. Even though there are thousands of different types of contract forms in use by contractors, there is a great deal of basic information that is common to all and that you want to be certain is included in your contract.

Obviously, there will be variations in the language depending on your situation and your particular project. Contracts vary from state to state because of differing laws and statutes, and there may even be differences in wording among areas within the same state in order to reflect such local variables as weather, soil conditions, and common trade practices. But these basic clauses and stipulations need to be in the contract somewhere, no matter how they're worded and what other provisions that contract may contain.

Basic Information

First of all, the contract needs to contain all the basic information relating to you and the contractor, including

> the contractor's company name, address, and telephone number
> the contractor's license number
> your name and address
> the address of the job, if different from your home address
> the date the contract was written

A Complete Description of the Work

This is perhaps the single most important aspect of the overall contract, since it describes the actual work to be performed and is the reason for the contract's existing in the first place. It is also the area where homeowners are typically the most lax, accepting only vague wording and descriptions of the upcoming project.

It may be that trying to insist on fuller descriptions of the work to be done is too intimidating for some people—involving as it does a number of early decisions on design and materials and some knowledge and experience in construction that you may not have—or it may be that it's simply too much trouble. Many people take the attitude that the contractor knows what he or she is doing and knows what you want, so why take the trouble to spell it all out.

Actually, there's some truth in that reasoning. There needs to be an atmosphere of trust established with the contractor, and at some point you'll need to accept the contractor's knowledge and skill and put the project in his or her hands—that's why you hired the contractor in the first place. You cannot specify each nail and each board to be used, nor can you specify the exact methods of installation.

The contract will typically contain the phrase "Work to be completed in a workmanlike manner." This is a somewhat vague description, although it has been tested in the courts often enough that "workmanlike manner" does imply intended compliance with some established standards under the law. A somewhat better description might be "Work to be done in compliance with all applicable local, state, and national codes and standards"; realistically, that will have to be your guarantee of the contractor's intention to erect the structure correctly.

So with that in mind, you don't need to get so specific in the contract that you are spelling out what size nails will be used in what area. If you are having the work inspected by the local building department, the building inspector will check a number of the more technical aspects of the work to ensure safety and compliance with the codes. (However, it should be pointed out that building inspectors don't catch everything and in most instances can't be held legally liable for anything they miss.)

What you do need to get in writing are the specifics of the agreement with regard to what labor is to be performed and what materials are to be used. Anything that was discussed *and agreed to* during your meetings with the contractor needs to go into the contract, including such things as "We'll go ahead and haul everything off for you" or "We'll take care of repairing that broken door while we're working in the kitchen." If it was specified and agreed to, and if you're paying for it, *put it in the contract.*

Material specifications are the area in which most people have a problem, either because they don't yet know what materials they want to use—oak or mahogany moldings for example, or painted walls in the dining room as opposed to wallpaper— or they don't know the options or the technical terms for the materials they have in mind.

You should have put enough thought into the project to have as clear an idea as possible about what you're going to do and how it's going to look (see Chapter 1). Remember: *If you don't know what you want, you can't expect the contractor to know.*

You also need to do some homework. If you were buying a new car, you'd look at the colors and the engines and the fabrics and the radios available, and you'd know how to answer when the salesperson asked you about options. The same holds true for a room addition or a remodeling: When the contractor starts talking about what kind of doors or windows you could use, you should have some idea what he or she is talking about. Forearming yourself with at least some basic knowledge will make you a more active participant in the project and will greatly lessen the chances that you will be given substandard materials or be otherwise taken advantage of.

Using Specification Sheets

If you have been given a written estimate for the work during the bidding process, most contractors will simply use those specifications for contract purposes as a description of the work to be done. This eliminates the need for them to have to repeat the same things again on the actual contract, which usually has limited space for a full description of the job specifications. This is fine so long as the initial description you were given is complete enough. If it was not, you will need to amend it accordingly before allowing it to become the full description of work for contract purposes.

Also, the contract needs to refer to that specification sheet, with wording such as "scope of work and specifications of materials to be as described in estimate dated

10/10/90, a signed copy of which is attached to this contract." Be certain that (1) the estimate sheet is identical to the one you were originally given (compare it carefully to your original copy); (2) the estimate sheet is signed and dated by the contractor; and (3) the date of the attached estimate copy matches the date specified in the contract (the contractor may have revised his estimate to you on different occasions to reflect changes in the specifications, so be certain that the most recent, fully amended copy is the one to which the contract refers).

However it's done—as a description right on the contract or as a separate sheet—the specifications need to be clear and complete. If you do not feel that the description of work to be done adequately reflects your agreement with your contractor, ask that it be changed accordingly. Don't sign the contract until you are comfortable that it reflects your complete agreement with the contractor. Here are some examples of how things should—and shouldn't—be worded:

Specify the Plans.   If the contract refers to plans—any plans, whether they are original pencil drawings, blueprints, architectural or designer plans, drawings or diagrams prepared by the homeowner or the contractor, or stock plans (fully prepared house plans purchased from a designer or mail-order design firm, as opposed to custom-prepared plans)—then at least once in the wording of the contract those plans need to be specified and referenced in a clear, unmistakable manner.

An example of the wording might be "As per attached [or "As per accompanying"] plans, prepared by the architectural firm of Smith, Johnson, and Jones, last revision dated 9/9/92." This clearly states whose plans are being used—an easy means of identification and an additional source of protection in a case where more than one person or firm submitted plans for the project. By specifying the date of the plans, you are assured that only the proper set is being used by all parties concerned.

After this reference has been entered once in the contract, the phrase "as per attached plans" or "as per specified plans" is then sufficient. This phrase, however, is *not* sufficient as an initial description, since it leaves the door open for a substitution of plans or confusion as to whose plans or which version of the plans are to be used.

This procedure is also used for specifying an estimate, bid sheet, or specification sheet that is being referred to in the contract. After establishment of the specific details in the contract, it can then also be referred to "as per attached specifications," or words to that effect.

Specifically Describe the Materials.   It's simply not enough for the specifications to read "ten doors." While this clearly states the quantity to be supplied, it gives no other details. The specifications should read something like: "ten doors, five 2–6 and five 3–0, to be oak solid core flush, with oak jambs and Colonial casings." This clearly lists the total number of doors (ten), how many of each size (five 2–6—meaning two feet six inches wide—and five 3–0), the material (oak), the construction (solid core), the style (flush), as well as the style and materials to be used in the jambs (the door frame—oak) and the casings (the trim around the door—oak in the Colonial style, which is a standard molding pattern).

An exception to this would be if a set of plans has been prepared with a listing of finish materials—called a "schedule." If the door schedule lists all the details—including size, type of wood, style of door—then it would be acceptable for the contract to simply read "ten doors, as per attached and specified plans."

Moreover, the specifications for doors should not simply read "with knobs." There are literally hundreds of knobs available, from those costing a few dollars to those costing hundreds. If you have selected knobs, be certain they are specified: "With ten Acme passage knobs [meaning they are nonlockable], 'Plymouth' model, in polished brass." If you have not selected knobs but wish to do so yourself later, the bid should have an allowance in it: "With ten knobs to be selected later by owner. Knob allowance $150." If you will select *and supply* the knobs later, the contract might read: "With installation of ten knobs, labor only. Knobs to be selected and supplied by owner."

Whatever you can specify at this point, do so: brand names, model numbers, sizes, colors, patterns, accessories—anything and everything. The main point is simple: Get it in writing.

Understand the Specifications.   It will often happen that you may not know what you should be specifying, and this is where your homework and common sense come into play. For example, suppose you were having a room addition done, and the contract or the specifications read "insulate attic." That may, on first reading, sound fine, but think again. How much insulation? What type of insulation? Which attic—the one over the room addition, the existing one, or both?

You need to know exactly how much insulation you are buying as part of this room addition package. The bid may read "Insulate attic with fiberglass insulation, as per code," which is an abbreviated specification saying that the contractor intends to do what is required by the building code and nothing more. This is fairly normal, and unless you have requested something different, it's acceptable from the point of describing the work. The only problem is you need to know what the applicable building code is, which means having to call the building department to request the information.

A preferable alternative would be if the contractor spelled it out for you: "Insulate room addition with blown fiberglass insulation to a level equaling R-30, as per code." This tells you the type of insulation (fiberglass), the installation method (blown), the amount (R-30, which is a standard rating for insulation levels), and also the added assurance that you are receiving a level of insulation adequate enough to satisfy the requirements of the building codes.

If your contractor intends to insulate the new attic over the room addition only, the contract should state that clearly so there is no misunderstanding. If, during the course of one of your conversations, the contractor mentioned that the existing attic could be upgraded while the room addition is being insulated, then your contract should state the terms of that agreement: "Insulate room addition attic to R-30 using blown fiberglass insulation, installed as per code. Upgrade existing attic insulation in rest of attic areas from R-11 to R-30, using blown fiberglass insulation as per code."

Specify Anything You're Doing Yourself.   If you intend to do any of the work on the project yourself, it's important that the details be fully specified to avoid confusion. This clause actually protects the contractor more than it does you, but it's important that everything be down in writing. The more you agree on and specify at the outset, the less likely it will be that problems or disagreements will occur later.

It's important to remember that your participation in this job in essence makes you one of the subcontractors, with a schedule to keep and the responsibility to do the job correctly and in a timely manner. The contract should specify what you are to do, when you are to do it, and how much time you have to complete your portion. Typical wording would be: "Owner to provide all necessary materials and labor for interior painting. Work to commence upon notification by contractor that interior is ready for painting, and owner will have ten days from that time in which to complete the work."

Some contractors, reluctant to work with an owner due to past problems, may add, for example: "Contractor not responsible for quality of materials or condition of finished product; nor for problems to the paint job or any other aspect of the work arising from improper paint or painting; nor for delays in the completion of the job or any costs associated with those delays that arise as a result of the owner's performance of his or her work." It's strong language, but many contractors have been burned in the past by homeowners who undertake work on the project that they don't have the skill or the time for, then blame the contractor for the finished product or for delays in the completion of the overall job.

Beginning and Completion Dates

Your contract needs to specifically state the dates when the project will begin and when it will be completed. These dates are important for your own planning and may also affect the amount and the timing of the contractor's payments, depending on how the payment schedules are prepared. Therefore, they need to be considered with care.

Time delays are often an area of disagreement between contractors and homeowners, with the contractors bemoaning the fact that homeowners don't understand the normal delays and problems encountered on a typical construction job and home-owners taking the attitude that the contractor is delaying the job on purpose to allow time to work somewhere else. There's often some truth to both arguments.

When you first start talking with a contractor, the second most common question (after "How much will it cost") is "How long will it take?" This is followed closely by "When can you start?" All contractors, no matter how long they've been in business, seem to be notoriously bad—or overly optimistic, to give them the benefit of the doubt—estimators of how long projects will take.

The contractor will size up the project, weigh the work to be done, estimate the number of subcontractors to be scheduled, factor in an allowance for the possible delays they might encounter, and then come up with an answer that invariably seems to be two to four weeks short for the average project—and escalates if the project is larger. The homeowner absorbs that time estimate—if the contractor says six to eight weeks, all the homeowner hears is six—and schedules the new furniture to arrive even before the paint is dry, followed closely by a hundred out-of-town guests for a family reunion.

There is no deceit intended. The contractor is trying to keep you happy and trying to sell the job. No matter what problems were encountered on the last job, he or she is always sure that the next one will run smoothly, so he or she is usually too optimistic. Just as you can be certain there'll be changes on a job, there'll be delays, also. If you can mentally prepare yourself for delays and be realistic and patient about them, you'll save yourself a lot of aggravation.

Start Dates.   Nothing would seem simpler than a start date as far as the contract is concerned. There will be a blank space—something to the effect of "work to commence on _____"—and the contractor inserts a mutually agreed-upon date. But then that date comes and goes, and the only progress the homeowner sees is a piece of equipment parked in the yard or a small pile of lumber blocking the driveway.

This brings up the question of what constitutes "commencement," which always seems to be a matter of perception. To the homeowner who has been waiting two years to be able to afford the project and two months more to get on the contractor's schedule, commencement seems as if it should be two concrete trucks and a crew of ten carpenters. To the contractor, yours might be one of twenty or thirty such jobs this year—you're on the schedule for a block of time beginning on that day, workers and subcontractors have been arranged for, materials have been ordered, and, to the contractor, the job is under way.

It might be wise to discuss with your contractor what will happen on the job as the work begins and progresses. You'll be anxious to see progress as quickly as possible, and while that's understandable, it might not be totally realistic. Ask what to expect as of that start date, and then prepare yourself accordingly.

Completion Dates.   Of more importance is when the job will be completed. This is the date you'll be anticipating for moving in, scheduling the delivery of new furniture, or just having the workers out of your hair after two months of noise and dust. It's a date you'll probably want to plan around with some degree of certainty, so it becomes an important part of the contract.

Most contracts have wording similar to "The approximate date the work will be substantially completed will be _____." By using the terms *approximate* and *substantially completed,* the contractor allows for some necessary leeway in setting the date when the job will actually be finished.

Establishing this bit of leeway is normal procedure and must be expected. There are a number of things beyond the contractor's control that can delay the completion of a construction project—weather, materials shortages or back-orders, illness. For this reason, the contractor will usually not allow the contract to specify a more exact completion date.

Contracts for very large commercial projects may include a penalty clause that assesses the contractor a certain amount of money per day in punitive penalties for each day the job exceeds the completion date. You can put a clause like this in your contract, but you will probably find that the contractor will actually decline the job rather than agree to those terms. Remember: It's in the contractor's own best interests to get the job done as quickly as possible so that the final check can be collected and the contractor can move on to other work; that alone is usually a better incentive than anything you'll get written into the contract.

On the other hand, there's no point in letting the wording of the contract give the contractor an excessive amount of flexibility in completing the work. Having a contract that states that the contractor is excused from performance for "acts of God and other conditions beyond the control of the contractor" may allow too many excuses for slow performance.

One question that is somewhat difficult to resolve is what constitutes "substantial completion." This term is obviously open to a variety of interpretations—the contractor may understand it to be when the building is weather-tight with all the utilities operational, while you may understand it to be when the house lacks only drapes and a mailbox. There are several ways of defining "completion" and "substantial completion," and you may want to amend your contract to include a better idea as to what you feel completion really is. Here are some suggestions:

• the issuance of a Completion Notice by city or county building officials
• the successful passing of all final inspections for building, electrical, plumbing, and mechanical systems
• the occupancy or use of the structure by the homeowner
• the abandonment of the structure, meaning essentially that the project has reached a point where the owner of the building clearly does not intend to

complete the work or occupy the building. This occasionally happens, particularly on the renovation of old or badly deteriorated structures. Most states have very specific laws governing "abandonment."

Future Changes or Modifications

Your contract needs to specifically address any changes that may be mutually agreed upon by you and the contractor. This includes changes in the beginning or completion dates, the scope of the labor to be performed, the type of materials being used, and the amount of the final price. Once again, put it in writing.

The contract will usually contain a clause stating that what the contract contains represents the complete agreement between you and the contractor. It should also state that any changes need to be in writing and signed by both parties. For example:

> Any alteration or deviation from the specifications set forth in this contract involving increased or reduced cost of materials or labor will be executed only upon written orders for same, signed by both the homeowner and the contractor, and will become an additional charge or a credit to the sums set forth in this agreement.

Never sign a contract with wording such as "I agree to pay for any and all additional work not set forth in this contract." That is simply giving the contractor a legally binding blank check to perform additional work that you may not need or want, with no prior notice to you and no agreement as to price or specification.

Most contractors use a simple change order form, which makes reference to the original contract and clearly states the nature of the change to be executed and the increase or decrease in the contracted price, if any. Some typical wording would be: "Substitute clear fir moldings for oak moldings in all rooms except living room. Credit to owner $500," or "Add one 4–0 × 4–0 double-pane sliding aluminum window in dining room, style to match existing window. Additional charge for labor and materials, $275."

Even if the proposed change does not increase or decrease the price, it's still important to have it in writing in order to avoid any questions or confusion. For example: "Change kitchen cabinet handles from Acme item 234 in polished brass to Acme item 432 in antique brass. No change in price."

Whether the contractor uses a change order form, makes notes on a separate piece of paper, or writes the change directly on the original contract, the same rules apply:

• Be sure all the details are written down and that they clearly reflect the entire scope of your conversation and agreement.

• Be certain the price increase or decrease is clearly stated. If the price change affects the payment schedule, that also needs to be in writing.

• Sign the change order to show that you have read it and agree with it, and be sure the contractor signs it also. If the contractor is simply amending the original contract, be certain that both of you initial the change on *both* originals.

• Obtain an *exact* copy of the change order for your files and keep it with the other paperwork relating to the job.

Stipulating the Price for the Job

The contract must reflect the full and complete agreed-upon price for the work to be done. Most contracts have two places where the price is to be entered. As it would be on a check, the price is written in as a numerical amount in one area and then spelled out on the line below.

Spelling out the amount in addition to writing it numerically is an old and simple precaution against the amount's being altered. No one could simply put a "1" in front of the numerical amount, since that would no longer agree with the wording of the written-out amount. If the contract has two spaces such as this, be certain that both are filled in completely and that the numbers agree. If there is only one space and it is filled in with the numerical amount, it's a good idea to spell out the amount at the bottom of the page and have the contractor initial it.

Setting Forth the Payment Schedule

Just as important as the amount is the time schedule in which that money is to be paid to the contractor. For a small job—say, $2,000 or less—the contractor may simply be paid when the job is completed. On larger jobs involving more money, the contractor will insist on being paid as the work progresses. The contractor will need a flow of money over the course of the job in order to keep employees, subcontractors, and materials accounts paid on time.

Remodeling.   For a remodeling project, it is typical to tie the payments to the completion of specific aspects of the job. This allows the contractor to stay current with expenses while ensuring that you are never in a position of paying in advance for work that has not yet been done. This is also an important consideration in protecting yourself against construction liens (see Chapter 5).

Here is a typical schedule of payments—called "draws"—for a $20,000 room addition:

Payment shall be made according to the following schedule:
1. $2,000   due upon signing of contract
2. $3,000   due upon completion of foundation
3. $9,000   due upon completion of rough building, plumbing, and electrical inspections
4. $4,000   due upon completion of drywall
5. $2,000   due upon completion of job

Note that the payments are specified as being due upon the *completion* of each phase, so you know that the work has been completed before you make payment. You will also notice that the draw amounts vary according to which phase has been completed. These amounts are usually requested by the contractor and reflect the amount of money invested in the project to that point. In this example, the third draw is the largest, since the contractor will have paid out quite a bit of money between the completion of the foundation—the point of the second draw—and the completion of all the rough inspections.

When reviewing a payment schedule such as this one, be certain that you're satisfied with the wording that sets forth the conditions describing when the draw is due, and also that you're comfortable with the amount of money being requested.

The first draw, for example, is actually a deposit or down payment. A few states limit the amount that a contractor can charge for his deposit—California, for instance, limits the deposit to $1,000 or 10 percent of the contract amount, whichever is less—while other states simply require that the deposit amount be clearly spelled out in the contract. While the contractor will make the request for a deposit, how much you are willing to pay is usually negotiable. *Do not* allow the contractor to talk you into an excessive deposit before work begins, no matter what discounts or other inducements are offered.

The amounts of the intermediate draws should be in line with the amount of work done, as shown in this sample schedule. Visualize the approximate amount of labor and materials that will have gone into the project at each requested draw point and see if that seems reasonable to you. For example, requesting that 50 percent of the job be paid for by the time the foundation is poured is out of proportion with how much work will have been done to that point—essentially, it's a way for the contractor to increase the amount of the deposit.

The real sticking point in any schedule of payments is what constitutes "upon completion of job." As with the phrase "significantly completed" when establishing a completion date for the whole project, there are many interpretations of when the job can be considered "completed." You may wish to establish with your contractor exactly what will be considered completion, and add that to the contract.

Hold-back Clauses.   You will want to retain a small amount of money—usually 5 to 10 percent of the contract price—as a final payment, to be paid when the job is completed to your satisfaction. The power of this final payment needs to be wielded honestly and with discretion. It should be used as an inducement to get your contractor back to finish up those few nagging odds and ends that always seem to haunt the end of a job. But if you're happy with the contractor and he or she has been fair and honest with you throughout the job, then the money shouldn't be withheld until every speck of dust has been cleaned and every five-cent part has been installed.

If desired, you can include a hold-back clause in the contract, stating that you will hold back money at the end of the job until satisfactory completion. A typical hold-back clause would be worded as follows:

> Final payment shall be due within _____ days from completion of the job. [You can define "completion" here if desired, or reference that section of the contract where it has already been defined.] Final payment can be withheld on account of defective work that is not remedied, subcontractor or material liens that have not been satisfied, or failure on the part of the contractor to make proper payments for labor, materials, or subcontractors.

New Construction.   If you are having a new home built, the draw system usually works differently. Typically, you will have a lender involved who has given you a construction loan to pay for the work. At the end of each month, the contractor will present you or the bank with a payment request—usually on the bank's forms—that will allow him or her to pay for materials and labor used to that point. If, for example, the foundation has been poured and the framing lumber has been delivered to the job site, those are the items for which the contractor will request payment. The bank will send someone out to verify that the foundation has indeed been poured and the lumber is on the site, then the bank will issue a check to you, the contractor, or jointly to both of you, depending on how the loan is set up.

The draw system actually works very well for everyone concerned. The bank physically verifies that the work has been done before issuing the check, which keeps its exposure on the loan limited to the value of the work actually performed. The bank's inspection protects you also, especially if you currently live out of the area and are not available to continually check on progress. And since you will be making an interest-only payment, based on the actual amount of loan money you have drawn at the end of each month, you're assured that the amount you're paying interest on is fair and accurate. Finally, the contractor is assured of being paid in a timely manner so that his or her payroll can be met and the accounts paid on time (thus ensuring that materials discounts are taken).

When the contract is drawn up between you and the contractor, you will need to spell out the draw agreement. The exact wording will vary depending on the nature of the agreement, but it should clearly spell out all the terms you've discussed. For example:

> Contractor to be paid on the basis of monthly draw requests, with draws to be limited to the value of work performed to that point. Draw requests to be presented by the contractor directly to Main Street Bank by the twenty-fifth of each month. Payment to be made directly from the bank to the contractor no later than the fifth of the following month. All draw payments are subject to inspection and verification by Main Street Bank.

If you are using your own money to pay for the new house, you can set up your payment schedule in any manner that's comfortable for you and your contractor. Your bank can probably provide you with copies of their draw request forms, and you can have your contractor request a monthly progress draw from you as he or she would from the bank. Or, you can set up a draw schedule similar to the one used in the remodeling example, tying each payment to the completion of the various phases of work.

For additional information on payments and payment schedules see Chapter 5.

Time and Materials

Some construction projects are undertaken on a "time and materials" or "cost plus" basis. In this arrangement, you pay the contractor for the actual hours worked and the actual materials used, plus markup and profit (that's the "plus" in "cost plus"), rather than having a preestablished, total bid amount.

Cost plus arrangements are common on small jobs where there is not enough money involved to warrant the contractor coming out to bid (calling a plumber to fix a leaking shower, for example) and on repair or remodeling jobs where the exact cost is difficult to determine prior to commencement of the work (for example, insect or dry rot repairs, where much of the damage is concealed in the walls and cannot be accurately known until demolition work begins).

Cost plus arrangements are sometimes used on remodeling or new construction projects also. Some contractors may insist on this arrangement if they believe the project is not well planned and will require time-consuming, on-the-job revisions, or if unfamiliar materials and products will slow down the work. Some homeowners may request cost plus if they believe that by monitoring the jobs and verifying the hours worked and the materials used, they are better assured of a fair price.

If you are entering into a time and materials or cost plus arrangement, be sure the terms are discussed and agreed upon, and that everything is *in writing*. Your contract needs to specifically spell out:

• How much the contractor is to be paid and whether the rate is to be figured on a per-hour or per-day basis. If you are paying a per-day rate, you should establish what constitutes "a day"—if the contractor is on the job for only seven hours one day, is that a gray area that may cause disagreements?

• Is there a ceiling limit on the cost of the work? It is a very wise precaution to establish a specific dollar limit that the contractor cannot exceed without informing you and obtaining written authorization.

• What hours will actually be charged for? Will you be charged for the time the contractor spends picking up materials (normally yes, since this is time spent on your job) and scheduling (again, normally yes).

• If your job is outside the area, will you be charged for driving time and, if so, is it both ways? A common arrangement is to pay for the contractor's time for driving one way to the job. There may also be a mileage charge; if so, this needs to be noted.

• What is the percentage of markup on materials and labor?

• When will the contractor be paid? Will it be at the end of the job, at the end of specific time periods (once a week, every two weeks, etc.), or at the completion of specific phases of the work?

Insurance Repairs

Hiring a contractor to do work on your home that is being paid for by your insurance company—repairing fire or storm damage, for example—is a special situation. The insurance company has to approve the amount being spent and will issue the check, but, except in rare cases where the contractor is employed directly by the insurance company, you are the one hiring the contractor. You will execute the contract, and you'll be the one ultimately responsible for paying the contractor.

Once a dollar amount for the work has been established (see Chapter 3), the contractor will prepare a contract. In most respects, the contract will contain the same basic information as the contracts discussed above. Under the section "Description of work," the contract will usually refer to the insurance adjuster's scope of work, which contains a complete description of everything that the insurance company is covering. Be certain that the reference is exact ("Work to be performed as per adjuster's Scope of Work number 2468, dated 11/11/90, a copy of which is attached"), so that there is no confusion as to which document the contractor is referring.

Payment is usually issued in the form of a two- or three-party check, with your name, the contractor's name, and sometimes the name of the lending institution holding the mortgage on the property. Occasionally, as many as five or six names may be on the check, including the holders of the first and second mortgages on the property, any co-signers on any loan you have that used the property as collateral, or anyone else who the insurance company's records have shown to have a financial interest in the property.

Before the check can be deposited, all of the named parties need to endorse it, which can become an incredibly time-consuming process. To further complicate matters, insurance companies sometimes issue a form of draft as opposed to an actual check, which takes additional time to clear the bank before the funds are actually available for use.

Contractors who regularly do insurance work are well aware of all these financial complications and will usually have a stipulation or two in their contract. They may request that you be responsible for obtaining all the necessary signatures on the check, and that you do so within a reasonable amount of time following completion of the job—usually ten days. They will also typically request that your deductible amount be paid either prior to beginning the work or immediately upon completion.

Insurance companies usually issue only one check, and it will not be issued until the end of the job. On a large job, this can present financial problems for the contractor, who may insist that arrangements be made for partial payments as the work progresses. If this is the case, you may be able to arrange with the insurance company to issue more than one check, or you may be able to get the check as soon as the contractor has agreed to the adjuster's Scope of Work and signed the contract. If you can get the check up front, have all the parties including the contractor endorse it, then you can deposit it to your account and issue payments to the contractor as needed.

Another common occurrence with insurance work is that the homeowner will take the opportunity of asking the contractor on the job to do some upgrading or remodeling. For example, the insurance company may be paying to replace part of your roofing where it was blown off during a storm, but you'd like to have the complete roof redone, so you'll be paying the contractor personally for the difference. All of this needs to be carefully spelled out in the contract, with descriptions of the work to be done, the cost of what's being covered by the insurance company, the cost of that portion you're paying for, and how and when the contractor is to be paid.

Warranties

Any contractor should be willing to warranty the quality of his or her work to the homeowner. Most contractors agree that the homeowner should expect and receive at least a one-year warranty on work the contractor performs. A warranty clause should be included in the contract, along the lines of:

Contractor will immediately remedy any defects in workmanship or materials that appear during the course of construction. [Any reputable contractor is going to fix problems as they occur rather than hide them—and in reality, chances are you'd never know about them anyway—but it doesn't hurt to include this wording in the contract.] Contractor further warrants any work performed under this contract or under properly executed change orders to be free from defects in workmanship or materials for a period of one year from completion of the work. This warranty extends to labor and materials supplied by subcontractors as well.

In addition to this warranty from the contractor, many materials, fixtures, and appliances carry their own warranties directly from the manufacturer. Be sure that the contractor supplies you with copies of all manufacturers' warranties, in addition to any instruction sheets and parts lists included with the item.

City and County Requirements

With any building project, there are certain legal requirements that need to be met concerning the ordinances of the city or county in which the property is located. Your contract needs to address these issues and clarify whose responsibility they are.

Building Permits.   In simple terms, if any of the work being done on your home alters the structure or any of its components (wiring, plumbing, etc.), increases or decreases the amount of living space, or changes the usage of a room, you will be required to obtain a building permit. The cost of this permit will be included in your estimate and should be reflected in a phrase such as "including any and all city or county building permits."

It is best to let the contractor be responsible for obtaining the necessary permits. The contractor will know what information is required and can work through the red tape much faster than you can. Also, if you take out your own permits, you will be ultimately responsible for seeing that the construction meets all the applicable building codes. Your contract or the attached specifications should clearly state who is responsible for obtaining the permits and also who is responsible for scheduling the necessary inspections by the local building department as the work progresses.

Zoning.   Every house is located within a certain usage zone that is usually established by the city or county planning commission. This zoning restricts what the property can be used for, thus protecting the integrity of the neighborhood. It prevents an apartment building from being constructed in the middle of a neighborhood of single-family residences, or a gas station in a noncommercial area.

Some of the things that you propose for your home may violate the zoning ordinances. For example, the zoning where you live may prevent you from setting up a shop or office for your business or from constructing a small rental apartment over the garage. In your contract, you'll need to spell out who will be responsible for determining the exact zoning in your area, and whether or not your proposed construction is in compliance with it.

Typically, establishing zoning compliance is the responsibility of the contractor. But bear in mind that some building projects may require many hours of research and meetings with planning commission officials; the contractor will include charges in the bid to cover this time.

Setbacks. Setbacks are the legally established distances from the edges of your property to the perimeter of any structure constructed on that property. This is done to prevent the building of houses that virtually touch each other and to maintain open space around each house to enhance the look of the neighborhood.

Setbacks are typically set by the planning commission, but these are usually minimum standards set for the entire community. The original developers of a subdivision can, for example, supersede these requirements to create more space between the houses, or deeper front or back yards. The setbacks can be increased—but never decreased—as part of the covenants of the subdivision, requiring that any new home or addition built in the future in that subdivision will have to comply with these increased limits.

Setback compliance needs to be verified prior to construction. Once again, this compliance is usually the responsibility of the contractor and should be so stated in the contract.

Used Materials

In remodeling situations, the homeowner often asks the contractor either to save used materials or, in some instances, even to reuse them. Kitchen remodeling is a perfect example: The homeowner may want the contractor to reinstall the existing dishwasher or stove or to save the cabinets so that they can be sold or used in the garage. Anything you specifically want saved for your future use or sale should be noted in the contract.

Most contractors are willing to exercise *reasonable care* in the removal of old materials, but none will guarantee that they can be removed safely and completely intact. Furthermore, few contractors will guarantee the performance of used appliances, fixtures, or materials. They will probably insist on specific language in the contract denying any responsibility for the installation or performance of used items, and excluding these items from coverage under your one-year warranty.

Cleanup

For some reason, job site cleanup seems to become an issue on many construction projects. Needless to say, there is a considerable amount of dust and debris generated on any construction job, and it's the contractor's responsibility to keep the site clean and safe and to remove the debris at the end of the project. An exception to this would be if, as part of the original negotiation process, you agreed to do the job cleanup and debris removal. In this case, you need to spell out your agreement with the contractor (who will usually insist on it) and then be conscientious in keeping up your end of the bargain.

What constitutes "clean" to the homeowner and to the contractor are usually two different things. Don't expect the contractor to wash the windows and remove the labels and dust the furniture when he's done—unless specifically requested and paid for, that thorough a cleaning is usually not included. You may, however, want a clause in the contract regarding cleanup in order to ensure that it gets done to your satisfaction.

The standard phrase in the industry is "broom-clean condition," and you will probably find something similar to that in your contract. If you can imagine what a job site would look like after a thorough sweeping, you have a good idea of what's included in his cleanup: All debris will be removed from the site and disposed of (unless you have agreed to undertake this chore); the floors will be swept of all wood chips and sawdust; excess materials will be hauled away; and usually a general, albeit none too thorough, dusting of the area will be done.

A conscientious contractor will go a step or two further, although you won't find it spelled out anywhere. If you've had your kitchen remodeled, he or she will usually dust or vacuum out the insides of the cabinets. New fixtures such as bathtubs and sinks will usually be wiped off and spruced up for you. If you have installed new appliances—a new stove or dishwasher, for example—all the packing materials will be removed, all the racks and other parts will be in place, all the books and instructions supplied with the unit will be collected in one place, and the contractor will make certain that the unit is working properly. Good contractors will also take the time to explain the unit's operation to you and answer any questions you might have.

Arbitration Clause

Despite the best intentions and the most thorough precautions, the possibility exists that a dispute between you and the contractor might arise during or after construction. In this event, some provision should be made concerning how that dispute is to be resolved.

Arbitration—the process of presenting your dispute to an impartial third party for resolution—is one of the most common and effective ways of settling a disagreement. It is a voluntary and fairly informal procedure, and making use of it usually means you can avoid protracted and costly litigation.

It is a wise idea to include in your contract an arbitration clause similar to the following:

> In the event of a dispute between the parties to this contract, both parties agree to submit their claims to arbitration. Arbitration to be performed by _____,* and the decision of the arbitrator, as well as any damages awarded, shall be binding on both parties.

Acceptance of the Contract

When all the negotiating is done and the contract is written to your satisfaction, you'll be asked to read and sign it. There will be a section entitled "Acceptance," and it will have a description of the responsibilities that you assume by signing. Read this section carefully!

Essentially, this section of the contract will explain that by your signature you are confirming that you have read the contract, that you fully understand it, and that you agree to all the terms and conditions it contains. If more than one person owns the property (if you are married or you co-own the property with another party), the contract may require the signatures of all interested parties, or it may contain the phrase "jointly and severally," meaning that by your signature you are guaranteeing payment jointly or individually.

The acceptance section may also include the terms under which you would be considered in default of the contract; what late charges, if any, you will be required to pay if you are overdue on making a progress payment to the contractor; and what fees you might be subject to should legal action be required to enforce the contract.

A contractor who is financing the project for you will be required to provide you with a Truth-in-Lending statement that spells out what your payment will be, what the interest rate is, what the term of the loan is, and other details. You'll be asked to read and sign this and will be given a copy for your records.

The exact wording of the acceptance section will vary considerably from contract to contract. It may be simple or it may be so full of legal jargon that it's indecipherable. Again, like the other terms of the contract, if you don't understand any of it, have the contractor or an attorney explain it to you before you sign.

---

*In states having a Contractor's Licensing Board, the board will usually have an arbitration process available, and this should be specified in the contract. Lacking that, both parties should agree who the arbitrator will be—usually a representative from the American Arbitration Association. See Chapter 5.

Cooling-Off Period

If your contract with the contractor falls under the federal Truth-in-Lending statutes and you are using your home as collateral for the loan, you have the right to cancel the contract within three days of signing it, without penalty. Some states provide a similar cooling-off period for any home-improvement contract, allowing you time to review it in private and cancel if desired. Since your house ultimately backs your loan because a contractor can attach a lien if necessary, be sure to include a cooling-off period in every contract (see Chapter 5).

Liquidated Damages.   Your contract may contain a "liquidated damages" clause. This essentially states that if you cancel the contract after the cooling-off period expires, you will be subject to certain charges. If it exists, this clause should limit the amount of damages you can be charged to no more than 5 percent of the job, assuming no materials have been delivered and no work has been performed. Once work has begun or materials have arrived, trying to back out of the contract—except in cases where the contractor has violated its provisions—would constitute a breach and could subject you to legal action by the contractor.

To find out if you have a mandatory cooling-off period or if there is a limit or prohibition on liquidated damages in your state, contact the state attorney general's office.

5

Seeing the Project Through—
Insurance, Liens, and
Your Relationship
with Your Contractor

In a perfect world where everyone is honest, where every businessperson conducts business with competence and intelligence, and where accidents can never happen, there would be no need for a discussion of the topics contained in this chapter. In the world we live in, however, this information becomes a vital necessity.

Many states require the licensing of contractors, and with that licensing come requirements for bonding and insurance. Other states require that the contractor be registered with a state regulatory agency, and that agency may also set standards for contractor insurance and bonding. Even if you live in a state where no such requirements are in force, it is *essential* that you deal only with bonded and insured contractors and subcontractors.

People who are involved in an automobile accident with an uninsured motorist may have to pay their own medical bills, even at the cost of all their savings, or they may have their cars totaled with no way to replace them. Having problems on a job with an uninsured contractor is very similar. However, while you have no control over who happens to hit your car, you do have control over whom you choose to let work on your home. Hiring a person who's uninsured is inviting trouble.

Insurance

There are three basic forms of insurance that a contractor needs to carry to protect himself or herself and the client: liability, workers' compensation, and automobile—depending on the type of business and the kind of work that is performed. Some contractors carry other forms of insurance as a general part of doing business. Others take out specialty insurance only when working on certain types of jobs or when using certain types of equipment. The contractor's personal insurance—health, life, homeowner's, etc.—need not concern you at this time (it would in fact become an issue only in certain litigation situations).

When you have your first appointment with a new contractor, simply ask whether he or she is insured. If your state requires licensing or registration, a call to the state contractor's board will verify that the contractor's insurance is in effect and that it meets the state's minimum requirements. If there is no state or county board, ask the contractor for the name and telephone number of his or her insurance agent. A call to the agent will verify that the contractor is insured, although the agent is probably not at liberty to divulge the amount of coverage.

Liability Insurance

Liability insurance is the form of contractor insurance that is of the most concern to you as a client and homeowner. In general terms, a contractor's liability insurance protects you if the contractor should, in the course of working at your home, cause damage to your house or an injury to you or anyone at your home.

For example, a roofing contractor who is re-roofing your home has removed all the old shingles from the roof when an unseasonable and unusually heavy snowfall hits. By the time the snow has been cleared away and the roof is covered again, there has been substantial water leakage in your home; damages to the house and its contents total over $30,000. It is the contractor's liability insurance that would pay for the damage.

Now suppose for a moment that the contractor was insured for only $10,000 or, worse yet, not insured at all. Depending on the limits of your own insurance policies, you might end up being responsible for paying the cost of the repairs out of your own pocket.

The same would hold true if the contractor dropped a board off a roof that struck and injured someone or uprooted your septic system or your award-winning roses with a tractor. Assuming that the contractor is indeed the one who caused and is liable for the damages, all these things would be covered by the contractor's liability insurance—up to the limits of the policy.

A contractor's liability policy typically covers (again, up to the limits of the policy) such items as

- damage to buildings, outbuildings (detached garages, barns, etc.), permanently installed fixtures and machinery, and personal property
- personal injuries caused by negligence
- damage to work under construction and materials stored on the job site

- the cost of temporary measures needed to keep the damaged building secure and weather-tight; living expenses if injured party's home is rendered unfit
- the cost of debris removal and damage cleanup
- the cost of cleaning up pollutants or repairing certain types of environmental damage

There are hundreds of small-print exceptions to the coverage—for example, injuries to pets or livestock may not be covered, as well as the loss of cash, certain other valuables, and certain types of plantings and other landscaping. If that hundred-year-old oak tree in the backyard that you love so much were to be damaged in some way, the insurance company might declare it simply "a tree" and award $100 for you to buy a new sapling.

The amount of liability insurance that a contractor carries is often set by the state or county in which he or she operates—typically in the range of $100,000 to $200,000—and represents the minimum amount that is generally considered adequate. In this litigious age, however, many contractors choose to increase those coverage amounts to $500,000 or even more in order to provide themselves and their clients some additional protection.

Workers' Compensation Insurance

Workers' compensation insurance is required for the contractor's employees and in some instances even for the contractor himself. Again, you would want to check with the state contractor's board or other governing body to see who needs to carry workers' compensation.

Workers' compensation is designed to pay for the cost of treating any injuries that a worker might receive while on the job, as well as all or some portion of wages lost while recovering from the injury. While insurance of this type is, of course, designed primarily for the protection of the worker, it has the additional benefit of helping protect you from needless litigation. If a worker is injured on your premises and is not protected by an employer's policy, chances are that, lacking any other recourse, he or she might try to hold you responsible.

Automobile Insurance

If the contractor has company-owned cars or trucks, they need to be covered by automobile insurance. This will protect you in the event of an incident involving these vehicles. Otherwise, for example, if an accident occurred involving an uninsured company truck carrying materials to your job, an injured party could name you in a suit for damages. Chances are that the person could not win such a suit, but if you're sued, you face the costs of hiring an attorney to defend you.

Coverage and rates vary by state and insurance company, but they are fairly high for commercial vehicles. In general, the insurance will protect you if the contractor injures you or damages your property with a company-owned vehicle, or if an accident occurs while in transit to your job or while hauling materials, equipment, or workers to and from your job.

Bonding

A bond is essentially an amount of money pledged or held as an assurance of performance by the contractor. There are a number of different kinds of bonds, depending on the situation and the desired use.

Contractor's License Bond

Contractors are required by most states to obtain a license bond as one of the qualifications for being licensed. The bond may be in the form of a cash deposit, held by a bank in the name of the state agency that licenses and regulates the contractors. More commonly, because of the large amounts of money involved, the bond is instead provided by a bonding company, for which the contractor pays a monthly or yearly fee.

The dollar amount of a contractor's bond may vary according to the type of contractor he or she is and the volume of work that is performed. The range is typically from $2,500 to $10,000 or more. High-risk contractors, such as those involved with the construction of swimming pools or the moving of houses, may be required to carry an even larger bond.

The bond is designed to provide a legal recourse for you in the event of a substandard performance by the contractor when the completed work does not meet the requirements of the county or state building codes. It will, in certain instances, also protect you against a breach of contract by the contractor.

In order to "attach," or get the proceeds from, a contractor's bond, you need to make a claim with the state builder's board or the agency having jurisdiction over the actions of the contractor. The board will hear the case, consider the circumstances, and, if warranted, issue a judgment against the contractor. The proceeds from the bond, up to the bond's value and up to the judgment levied by the builder's board, will then be paid to you.

It's important to point out that the bond's value is an "aggregate amount," meaning that's all the money that is available to settle all the judgments against the contractor. If ten people file claims and receive judgments to attach the contractor's $10,000 bond, that $10,000 is divided among all the claimants.

Again, like liability insurance, the bond is extremely important to you when hiring a contractor. It's highly unlikely that you will ever need to go after a contractor's bond—or need to rely on his insurance—but it's essential that you verify the existence of both before signing a contract.

Finally, the license bond is a protection only against substandard work and certain types of breach of contract. It will not protect you against the contractor's insolvency or the contractor's failure to complete the work—for that type of protection, you would need a performance bond.

Performance Bond

While a license bond is required by most states, the use of a performance bond is optional. It is something that you will need to request and pay for separately. The performance bond guarantees that no matter what happens to the contractor, your project will be finished.

Performance bonds are very common on large commercial projects, although they are not seen on residential projects nearly as often. If your project is a large one financially, it may be worth looking into. The cost is typically around 2 to 3 percent of the cost of the job, so the performance bond for the construction of a new $100,000 house would run around $2,000 to $3,000. The expense is the main reason few people request them.

The performance bond ensures that funds will be available for the completion of your project, no matter what. For example, suppose you hire a contractor to construct a new home for you, and you agree upon a price of $100,000. Partway through the construction, the contractor stops work for illness, bankruptcy, or any of a number of circumstances and is unable or unwilling to complete the job. You have already paid $75,000 of the agreed-upon $100,000, even though the contractor has not really completed $75,000 worth of work. You contact another contractor who will charge an additional $40,000 to come in and finish the house, leaving you with a net loss of $15,000 by the time the job is completed.

If you have a performance bond in effect, the bonding company will either hire a contractor to finish the work according to the original plans and specifications or settle with you for a cash award.

If you wish the contractor to provide a performance bond, you will need to make this a stipulation during the bidding phase of the project so the contractor can allow for its cost in the bid. However, this may limit the number of contractors who will be able to bid your project, since the obtaining of a performance bond requires a long track record and a very stable financial background that a new, small company may not be able to satisfy.

Payment and Contract Bonds

Payment bonds and contract bonds are also not often seen in residential construction and remodeling, especially for small jobs. Both are optional bonds that your contractor can obtain if you stipulate your desire for one prior to the signing of the contract. Once again, you will be charged for the additional costs.

A payment bond assures you that no liens for labor and materials can be filed against the property. A contract bond guarantees you both job completion and the payment of all labor and materials liens. Both offer excellent additional protection, but they may be too costly and actually unnecessary in most instances.

Construction Liens

Construction liens represent an area of the law of which consumers are unaware, or do not fully understand. Lien laws vary widely from state to state, and you should take the time to find out about them before hiring a contractor.

Essentially, the laws in most states provide that people who furnish labor or materials to your home during a construction project are entitled to file a construction lien (sometimes called a mechanic's lien) against your property in the event they are not paid. This is the case even if you have paid your contractor in good faith. If your contractor has not paid the subcontractors, you are in a potential "pay twice" situation.

For example, say you've paid your builder $20,000 during the course of construction. You have seen materials being delivered to your house. Subcontractors have come and gone, many of whom you've never met or have seen only once or twice. You naturally assume that since you have made your own regular payments, the contractor has used the money to pay his or her material and labor bills.

At the end of the job you find that this is not the case. The contractor has not been paying bills, and the subs and the material suppliers now turn to you for payment under the provisions of the lien laws. Of the $20,000 you paid out, only $5,000 was used to pay bills. Another $9,000 is still owed. If the subcontractor or material supplier followed the procedures correctly, you could find yourself with a $9,000 lien against your property.

If you are unable to pay the additional $9,000 to satisfy the bills—again bearing in mind that you have *already* paid the $20,000 of the original contract—foreclosure proceedings can be initiated against your home, which can be sold to generate the necessary money. Fair or not, this is all perfectly legal. It has happened to a number of homeowners.

Contractor Lien Information Notices

In many states, the contractor is required by law to provide you with some sort of lien information prior to signing the contract. This is done to inform you about what liens are and how they affect you and to avoid confusion and concern arising from the receipt of a preliminary lien notice.

For example, Oregon requires that contractors provide a form entitled Information Notice to Owners About Construction Liens if the project contracted for exceeds $1,000 in value. The notice states, in part:

> If your contractor fails to pay subcontractors, material suppliers or laborers or neglects to make other legally required payments, those people who are owed money can look to your property for payment, even if you have paid your contractor in full. This is true if you: have hired a contractor to build you a new home; are buying a newly built home; are remodeling or improving your property.

This notification goes on to explain another area of the lien laws about which even fewer people are aware, concerning your purchase of a newly constructed house even if you did not hire the contractor who built it:

> If you enter into a contract to buy a newly built home or a partly built home, you may not receive a Notice of the Right to Lien [Oregon's name for a preliminary lien notice]. Be aware that a lien may be claimed even though you have not received notice.

Each state has different laws and requirements governing the filing, release, and enforcement of liens, so ask your contractor or the state contractor's board for more information. The Oregon notice sums up your liability rather succinctly:

> You have final responsibility for seeing that all bills are paid even if you have paid your contractor in full.

Preliminary Lien Notices

In most states, subcontractors and material suppliers are required to send you a preliminary lien notice within a certain number of days after beginning to supply labor or materials on your job. Failure to comply with these lien notification regulations will result in that person's loss of the right to file a lien at a future date.

Shortly after work begins, therefore, you will probably begin to receive these notices in the mail. Don't become alarmed, but don't ignore them either. While the notice does not mean that a lien has been filed against you, it does mean that someone is supplying labor or materials to your home and is notifying you of his or her *right* to lien.

Keep these preliminary lien notices in your job file, and let the contractor know about them as they arrive. As the job progresses and you make payments to the contractor, these will give you a record of who should be getting paid with your money.

Lien Releases

As your job progresses and you make the necessary payments to the contractor, request that you be supplied with lien releases for anyone who has sent you a preliminary lien notice. A lien release is simply a form signed by the subcontractor or material supplier stating that payment in full has been received and that he or she is releasing the claim of lien against your property.

A lien release is the only way you can be certain that the person has been paid. In your contract, it is a good idea to tie your final payment in with the obtaining of all necessary lien releases by the contractor. If there is a lender providing construction money to the contractor, the lender will usually stipulate that lien releases be obtained at various stages of the project before it will provide any additional funding.

Funding Control for Your Construction Project

One way of helping to prevent liens and other problems is to have your contract stipulate some sort of funding control. Funding control is exactly that: having some sort of additional control over paying the contractor than merely writing a check. There are a few options worth considering, but remember that whatever you choose to do, the details need to be worked out in advance with the contractor and specified in your contract.

Funding Control Companies

Use of a funding control company is one very good way of handling your payments to the contractor—in fact, many lending institutions will require that contractor disbursements be handled through an independent funding control company in order to protect their interests as well as your own. Even if you do not have a lender, you might want to consider a funding control company.

A funding control company is essentially an escrow company that specializes in construction projects. You or the bank—whoever is providing the money—pays directly into the funding company account. It in turn takes over control of disbursing the funds to the contractor, subcontractor, and material suppliers, according to your specific instructions. It will also obtain the necessary lien releases as the job progresses and the bills are paid.

The funding control company offers you the advantage of acting as a central processing point for all the paperwork generated on a large construction project. It makes certain that the bills and payment requests are in order, and that the payments to the contractor never exceed the actual value of the labor performed or the materials supplied up to that point.

If you are considering the use of a funding control company, ask your bank for suggestions and recommendations. Talk with the control company, find out what services it provides and what its fee is—usually a small percentage of the overall cost of the construction project. Some companies will make on-site inspections to determine that the labor and materials for which the contractor is requesting funding have actually been used on the job. Others simply request that you fill out and sign a voucher, authorizing payment to the contractor.

Subcontractor Statements

If you are paying your contractor directly, you can request that each subcontractor or material supplier who sent you a preliminary lien notice provide a current statement. This will let you know exactly what is currently outstanding against the contractor's account as it applies to the work being done on your property.

The request should be made in writing and should be sent certified mail with a return receipt requested. The subcontractor or material supplier is required by law in most states to respond, usually within fifteen days.

Two-Party Checks

You can use a two-party check, made out to both the contractor and the subcontractor or material supplier for the actual amount of the bill that's due. This ensures that the contractor cannot use the money for anything other than what you intended it for, but it does present some complications for the contractor. Be sure to discuss this option with the contractor before signing the contract.

Maintaining a Smooth Relationship

For the person who's waited years to build a new house, to remodel a kitchen, or to add on a family room, nothing quite compares with the sight of a lumber pile in the driveway and a group of carpenters showing up bright and early one morning. There are dreams of what the new spaces will look like and excitement about reviewing the work daily to see what new things have been created.

Yet there is also nothing that compares with living through the reality of a construction project, with the mess, the delays, the decisions, the arguments, and the disappointment of days of inactivity.

The precautions you have taken up to this point will be a tremendous help in establishing a solid working relationship with a reputable contractor. And now that the job is under way, there are a number of other things you can do to keep the relationship smooth and the project exciting.

The Construction Process

Like any relationship, understanding a little of what the other person is doing can help you get through a number of small rough spots. The same is true for your construction process: The more you know about what's going on, the better prepared you'll be for it as it happens.

This is not to say that you need to take a course in carpentry or construction management. You don't need to know exactly how two boards go together, but you certainly should know something about what to expect when the crews arrive and the hammers start swinging.

The Mess

First and foremost, keep in mind that the construction process is a messy one. There is dust, there is noise, and there is a daily accumulation of debris of all shapes and sizes. This is perfectly obvious to anyone who has ever seen a construction site, but the average homeowner is shocked and dismayed when it happens in his or her house.

The mess is exceptionally difficult during remodeling, especially if you're living in the house as the work is going on. The contractor may put down tarps and hang plastic in an attempt to protect things against damage as much as possible, but these precautions soon seem futile. Dust filters in around masking tape to coat the furniture, dirt gets tracked around tarps and down to the phone or the bathroom that the workers use, and the dog delights in bringing you a fresh collection of wood scraps each evening.

You need to remind yourself that this is an unavoidable fact of the remodeling process. It can be mitigated to some degree, but it can't be eliminated. The sooner you accept that your home will not be totally spotless and organized until the work is finished, the easier it will be on your nerves.

To some homeowners, the mess is not a problem. Either they are used to a little clutter in their daily lives or they're so happy to see progress on the job they can accept the inconvenience of a little mess.

To others, this proves to be a daily battle of patience and adjustment. If you feel you will have a hard time living with the mess, talk to your contractor about it in advance. Perhaps a more thorough daily cleanup than the "broom clean" specified in the contract can be provided, but you will be charged extra for this service. You might also want to consider hiring a housecleaner or janitorial service to come in on a regular basis in order to help keep your house in order.

Some people arrange not to be around at all during the remodeling. They may choose to schedule their vacation at that time, stay with friends or relatives, or rent a motel room. One homeowner went so far as to rent another house right down the street so that he could visit the site and oversee the work without having to live with the mess. How far you want to go in the quest for cleanliness and order is up to you and your family, but it's certainly something that should be discussed and considered before the work actually begins.

Living with a Partial House

The other thing you can count on during the process of remodeling or repairing a house is that part of the house will not be usable. Like the mess, this is another perfectly obvious result of construction that nevertheless comes as quite a shock to the family who has to live through it.

There will probably be, at various times and for various durations, no electricity, no gas, no heat, and no water. You may be without a sewer system for a short period, and an operating toilet may be a luxury you're denied for a day or two. There may be a plastic-covered hole where the window used to be, or a sheet of plywood to serve as a temporary front door.

Adjusting to life with a partial house is more a matter of patience and preplanning than anything else. If you are losing a bedroom, for example, don't wait until the night before the job starts to figure out where you or the kids are going to sleep. Make arrangements to double up in another bedroom, or rent a rollaway bed for a while and set it up where it will be out of the way.

Kitchen Remodeling.   The biggest shock to the household system is the loss of the kitchen. As cooking center, social center, and nerve center of the entire house, the kitchen is difficult to lose use of. If you are planning to have your kitchen remodeled, here are some tips:

• Ask your contractor for as accurate a schedule of events as possible. That way you can plan accordingly for meals out, social events, and even holiday dinners.

• A few days in advance of the demolition work, start setting up a temporary kitchen somewhere else in the house, such as a spare bedroom or a corner of the dining room. You can move the refrigerator there and set up a microwave. If you don't currently have a microwave but are getting one as part of the remodeling, ask the contractor if it is possible to have it delivered earlier, so you can use it. You can also set up a camp stove for temporary cooking, but *be sure* you provide adequate ventilation. Do not, under any circumstances, use a barbecue or other charcoal grill indoors—the fumes are deadly in enclosed places.

• A few large cardboard or wooden boxes can be used as temporary cabinets to store food, with smaller boxes for plates and utensils. Set a sheet of plywood on the boxes for a counter, or use card tables.

• The night before demolition is to begin, empty the kitchen. Place the contents of the cabinets in boxes. Store the boxes somewhere safe and out of the way of the work crews, and cover them to keep the dust to a minimum. Pack away canned and dry goods and other items that won't spoil. Keep out only those utensils and food items that are essential.

• Keep your meals simple—cereal for breakfast, canned soup and sandwiches for lunch, microwavable frozen dinners—or plan on eating out. Set up an area in the bathroom to wash dishes. Buy paper plates and cups and plastic silverware to keep dishwashing to a minimum.

Bathroom Remodeling. Second only to the kitchen in inconvenience is the loss of a bathroom. If you have a two-bathroom house, plan in advance to move everything out of the bath being worked on and set up your toilet articles in the second bathroom.

If you only have one bathroom, you'll need to do even more advance planning. Bathroom remodeling can easily take two to three weeks, depending on what's to be done, which can present some serious inconveniences. Your contractor can arrange to have a temporary outhouse brought in to serve your needs for toilet facilities. These portable toilets are weather-tight and private, and the rental companies come by on a regular basis to be certain they're pumped and washed to keep them sanitary.

If the job will be going on for a while and you'll be without a shower, arrange to use one at a friend's or neighbor's house. A camper or motor home in the driveway can also provide temporary facilities; these can be rented temporarily.

Room Additions. At least initially, room additions will cause less disruption to the daily routine. Most of the work is done outside and, other than the noise, shouldn't affect you too badly until it comes time to open up the house to the addition.

• Ask your contractor to construct as much of the addition as possible, put on the roof, and make sure the room is weather-tight before opening up the new doorway or removing the adjoining wall to the existing house. This will keep dirt, noise, and cold to a minimum and also keep your house more secure.

• Request that material and debris piles be placed in areas where they have as minimal an impact on your daily life as possible. You might want to clear out areas in the backyard, garage, or driveway for the contractor to use for storage and debris, in order to maintain a little more control over where things are dumped. Make these areas convenient to where the work is being done.

Cleaning Out the Job Site

One thing that is often overlooked during the planning and negotiating stages with the contractor is who will be responsible for cleaning out the area where the work is to be done. Emptying the kitchen cabinets, for example, or moving all the accumulated items out of a garage that is going to be worked on can be a formidable task.

Unless previously arranged, the contractor is usually not responsible for emptying rooms. The contractor will not want to spend time emptying shelves or moving clothes out of the closets, and will not want to assume the responsibility for packing and storing them safely.

Before the work is scheduled to begin, clean out the rooms involved. Store things in rooms that aren't being worked on or out in the garage or storage shed, or consider renting a mini-storage building for a few weeks. The contractor will usually be happy to help you move a heavy piece of furniture or two—as a courtesy. However, the contractor will expect any room in which work is to be done to be as empty as possible when the work crews arrive.

Maintaining a Job File

From the moment of the first glimmering of an idea, your construction project will generate a large amount of paperwork. From preliminary sketches to bills and receipts to manufacturer's warranties and instructions, there's a lot of important paperwork to keep track of. This is why you need a job file.

A job file is simply your own storage and filing system for paperwork related to the project. You need a central place to store everything, organized so you can find anything you need later. The job file can be several file folders, an alphabetical accordion file, or even a shoe box. The only criterion is that it works for you so that you'll use it regularly.

An inexpensive accordion folder seems to work best for most people. Available at stationery or office supply stores, these folders have pockets that are labeled A to Z or can be relabeled to suit your own filing system. Individual file folders will also work well, especially for smaller jobs.

Many people find that the following eight categories serve well as a basic filing system. You can always add categories as the need arises.

- Design ideas: During the course of your discussions with your family, the designers, and the contractors, you'll create a number of lists, notes, ideas, and other written memos. You'll also end up tearing pictures out of magazines and perhaps even photographing other homes in the area. Keep all of that in this file so you'll always have your ideas close at hand should anybody want to discuss them.

- Legal and financial papers: Keep a copy of the original estimates, the original bid, and the contract in this folder. You'll also want to keep copies of all your loan papers, disclosure statements, credit reports, and any forms or other papers related to the bank's inspections and the draws you've taken.

- Change orders: Keep this file exclusively for change orders. If you have requested a change, or if the contractor has discussed one with you but has not yet provided a written change order, write a note—with the date and time of the request or conversation and a description of the proposed change—to keep in this file until you have the contractor's written change order.

- Communications: As the job progresses, you will be having any number of conversations with your contractor and the subcontractors. While it's certainly not necessary to keep track of every word, it's often a good idea to keep a few notes as the job progresses. You might want to jot down a brief description of what was done each day, or any problems or minor changes that came up (keep your notes on *major* changes in the change order file). If the contractor tells you he or she is going to do something, make a note of it. This will be a good aid for your memory and will prove an invaluable resource should problems arise.

- Lien information: Some states require contractors and subcontractors to give written information about liens and to send out preliminary lien notices in order to protect their lien rights. Keep anything relating to liens in this file.

- Warranties: Many of the new appliances, fixtures, and even building materials that will be used on your job carry their own warranties directly from the manufacturer. Ask your contractor to save these for you or collect them yourself on the job site—they tend to get thrown away with the empty boxes as the items are installed. You can also use this file for manufacturer's instructions; however, they tend to be bulky, so you may want to file them elsewhere.

- Receipts and invoices: As the job moves along, you will find yourself with a number of receipts for materials that you purchase yourself, as well as copies of paid invoices from the contractor. Keep all of these together as proof of what you've bought and what bills you've paid; this will simplify matters should you need to return something. Keep this file indefinitely, for tax purposes—if you sell the house, many of these items can be deducted, offsetting your capital gains tax.

- Photographs: As the job progresses, you may want to take photographs of the work. While these are fun to review later for a "before and after" look at the project, their value really lies in the event of future problems. When taking the pictures, try not to get in the way of people as they're working, and don't give the impression that you are overseeing and documenting each move they make. You may even want to do your picture-taking after the crews have left for the day.

Handling Draws and Payments

Just as your contractor has a contractual agreement to perform certain work by a certain time, you have an obligation to pay him or her as per the schedule specified in your contract. Remember: The contractor has a number of obligations to meet with the payments from your job—payroll, material bills, etc.—and needs to be paid in a timely fashion. Most contracts have penalty stipulations in them, charging some sort of late fee on payments not made within a certain number of days following the due date.

Review your contract to know in advance when payments are going to be due. If the payment is tied to the completion of a particular phase of the job, ask the contractor when that is expected to happen. Have a check for the payment amount ready, plus or minus the cost of any changes to that point, and present it to the contractor as soon as that phase has been completed to your satisfaction.

If you are getting draws from a bank to pay for the work as it progresses, it is usually the contractor's responsibility to prepare the draw request. The contractor will fill out a bank form that specifically lists the work that has been completed or the materials that have been purchased since the time of the last draw and will make a request for an amount sufficient to cover those costs. Depending on the arrangement, the draw request may be submitted directly to the bank or it may go to you first.

After the bank has made its inspection and approved the draw, usually within two to three days of the request, a check will be issued. The check may go directly into your checking or savings account at the bank, it may be paid directly to the contractor, or it may be made out jointly to both you and the contractor—again, this depends on the policy of the bank you're dealing with and the specific arrangements you have made with the contractor.

The Final Payment

This is the moment that can present the greatest difficulties for both the contractor and the homeowner. The contractor believes the job is finished, perhaps with the exception of a few items that are back-ordered or awaiting an opening in a subcontractor's schedule. Therefore, the contractor wants to finalize all the paperwork, pick up the final payment, and move on to the next job.

On the other hand, the homeowner may fear that the moment the final payment is released, the contractor will be completely inaccessible. If the contractor never comes back to fix anything, those few odds and ends will remain eyesores and reminders forever, and the homeowner will berate him- or herself for releasing that check. The homeowner may also start feeling the need to closely examine the entire job, figuring that all claims must be made now.

Both the contractor and the homeowner have a point. The job is essentially done, so the contractor should not have to wait weeks or even months for the money owed simply because a two-dollar part is back-ordered from the factory. On the other hand, once the final payment is made, that two-dollar part will tend to languish on a shelf in the contractor's office until a little time is found to schedule someone to go install it. The contractor has no incentive, other than honesty (which unfortunately is not always enough), to finish the minor details.

There are two reasonable solutions to this dilemma. First, in your contract, agree to a specific period of time after completion before the final payment is due. It may be ten days if the contractor gets his or her way or thirty days if you get yours, but whatever it is, hang on to the payment for that period. This will give you a chance to live with the work, to make certain everything's functioning correctly, and to see if anything else comes up.

During this period, make a "punchlist"—that piece of paper contractors hate—that itemizes those things about the job you're not happy with. Be thorough, but *be reasonable*. Go over the list with your contractor, asking that those specific items be taken care of. That's fair to the contractor, who knows what has to be done and has had a chance to discuss it with you, and it's fair to you—you can feel comfortable that the job is done well and any major problems have been resolved. When the items on the punchlist are completed, you can issue the check and close out the job.

The second fair way of handling things is to withhold a small portion of the final payment rather than the entire final draw until the punchlist is done. If you are satisfied with the contractor and the work done, show that by honoring your obligation and paying as soon as the job is finished. If most of the repair items have been taken care of in good faith but you're still worrying about the final part or touch-up that he promised, pay all but $100 or $200 of the last draw until those things have been taken care of.

Constant Communication

As your job proceeds, stay in touch with your contractor about what's been done and what's going to happen next. While you don't want to be underfoot to the point that you're absorbing an inordinate amount of the contractor's time, it is your house and you deserve to have an idea of what's going on.

Staying in touch is also your best way to see what the actual, three-dimensional building looks like as it comes together. Most people cannot accurately visualize exactly how something is going to look just from a set of plans and may not be happy with the way something is turning out when they actually see it built. Obviously, you don't want to be tearing things out and changing them every five minutes, but if something's going to be done that you really don't like, it's better to discover it early, while it's still possible to fix it.

Discuss changes and problems as they come up. Remember: You may not always understand how a contractor is doing something, and a job may look disorganized and even a little shabby while still unfinished. However, waiting until the work is almost completed before pointing something out or asking a question—hoping that it will get corrected if you just ignore it—may create a major problem or a major repair. Contractors do make mistakes, or they may misunderstand how you want something done. Despite the best of efforts, only so much detail can be written into a contract, so don't be afraid to ask a question or request that something be changed or repaired.

If you suggest a change that is not due to contractor's error—for example, the plan called for a five-foot window but now that you see the opening framed, you'd really rather have one that's six feet—you should expect to pay for the change. However, by pointing it out early—in this case, during the framing stage—it's much easier and less expensive to change than after the window has been ordered and installed.

Take the time to discuss problems and new ideas with the contractor. Ask how and why things are being done a certain way if you feel that's necessary for you to understand them. Good lines of communication are the best way to avoid problems and hard feelings before they become major issues.

If Major Problems Arise

If a major dispute arises over how something is being done, sit down with the contractor (and any subcontractor who's involved) and discuss the problem calmly. Your architect or designer should also attend the meeting.

Point out what it is you're not satisfied with, and discuss what options are available to you. There may be design changes that can be made, or perhaps different materials or products can be used. Talk over how long it will take to remedy the problem, how much it will cost, and who will pay for it. A shouting match with accusations and threats of lawsuits won't get you anywhere—a calm, reasonable discussion usually will.

If necessary—although this is a last resort—ask that work on the job be halted until the problem is straightened out. If the problem is not the contractor's fault, you may be charged for the down time and the rescheduling, but again, it's better than having work proceed, thus risking irreversible damage.

Arbitration

If you do have problems that reach the point where one-on-one discussions between you and the contractor yield no solution, there are still options open to you before having to resort to legal action. The best of those is arbitration.

Many state contractor's licensing boards maintain an independent, impartial arbitration service to assist the homeowner—and the contractor—in disputes over contracted work. Arbitration is simply a chance for you and the contractor to sit down with a third party—the arbitrator—and explain both sides of the dispute without having to resort to the delay and expense of hiring attorneys and taking action through the courts.

The arbitration proceeding is simple and informal. No attorneys are used, no jury or other panel of people will be involved. There are no lengthy procedures or paper shuffling or vague legal terms and concepts beyond the comprehension of the average person. It's just you and the contractor, telling your own story in your own words and getting an impartial response.

As is so often the case, there is seldom a clear right and a clear wrong in disputes on a construction job. You wanted this and said this; the contractor thought you wanted that and said that. Assuming that both of you are acting in good faith and feel that you have been misunderstood or treated unfairly, arbitration offers a fair and inexpensive course of action.

The arbitrator will listen to a complete explanation of the dispute from both parties and will occasionally hear witnesses if the information they have is deemed crucial to the problem. The arbitrator will also review all the legal documents involved, including the contract, the bid, change orders, bank papers, and photographs of the job.

Within a very short period of time, usually no more than a couple of days, the arbitrator will present his or her findings. Making every effort to cut through the rancor and get to the essentials of the dispute, the arbitrator will offer practical solutions for resolving the dispute and finishing up the job.

Keep in mind that if there is an arbitration clause in the contract and one party chooses to initiate the arbitration proceedings, the other party is obligated to participate. Failure to join in the arbitration or the taking of other legal action of your own can constitute a breach of contract and seriously jeopardize your case, as well as possibly opening you up to other legal problems.

Binding Arbitration.   There is both binding and nonbinding arbitration, depending on the agreement you and the contractor enter into before beginning the arbitration proceeding and the laws governing arbitration in your state. While both types of arbitration are useful and are certainly better than shouting matches between you and the contractor, binding arbitration is preferable since it forces a conclusion to the dispute and dictates that action be taken.

With binding arbitration, both parties agree that the arbitrator's decision is final and that each is required to act on it.

For example, say the dispute was over an expensive change in the specifications on the job, a change the contractor made and for which you refuse to pay. The arbitrator may conclude that your actions indicated that you wanted to make the change and led the contractor to execute the change, but the contractor misunderstood the intent and scope of what you wanted changed and charged too much for it.

The arbitrator decides that you do have an obligation to pay for the work, but that the contractor needs to reduce the charges relating to these changes by half. If you have agreed to binding arbitration, you are both then obligated to accept this decision and proceed accordingly—the contractor to lower the bill, you to pay the reduced amount. Failure to act on the results of binding arbitration will almost certainly result in legal action by the other party.

Nonbinding Arbitration.   With nonbinding arbitration, both parties participate in the proceeding and listen to the result but are not bound to act on it. Perhaps you still don't agree that you should have to pay the bill, or the contractor still doesn't feel that he or she should have to lower the charge. Nonbinding arbitration cannot force either of you to take those actions.

At the least, nonbinding arbitration offers each party the opportunity to air his or her grievances in front of a disinterested third party. Talking over a dispute can often clear the air and lead to a compromise solution. However, be aware that when entering into nonbinding arbitration, a final resolution of the problem is not assured.

Beginning an Arbitration.   The procedures for initiating an arbitration proceeding are different in various states and jurisdictions. You can usually request an arbitration through the state builder's or contractor's board, which will have its own arbitrators or can arrange for them. Working through the contractor's board also has the advantage that the board, depending on the outcome of the proceedings, can initiate attachment of the contractor's bond.

You can also contact the state attorney general's office for assistance, or the Better Business Bureau, the local Chamber of Commerce, or any of the state or local consumer protection agencies—any of these groups should be able to help you find an arbitrator who is mutually agreeable to you and the contractor. You can also usually file a request directly with the American Arbitration Association or any other association or agency that provides arbitrators.

There may even be specific language in your contract that will force an arbitration in the event of certain disputes. If there is such a clause, it will usually be accompanied by a clearly defined set of conditions and procedures for initiating the proceedings. As a last resort, you might wish to contact your attorney, who may be able to arrange and begin an arbitration proceeding for you.

The Courts

When all attempts to resolve a dispute fail, your final recourse is through the courts. Court proceedings vary depending on the type of dispute and the type and amount of damages you are claiming. For a dispute involving a relatively small amount of money, you can file through small claims court and handle the case yourself without a lawyer. Larger or more involved claims will require the assistance of an attorney.

Remember: A court proceeding is rarely the dramatic triumph of justice that the television shows tend to portray. You will be undertaking a heavy financial burden by hiring an attorney—sometimes more than the monetary value of the dispute itself. Court proceedings are slow and tedious and may take years before reaching any kind of solution—years in which your construction project may be sitting unfinished. For your sake and the contractor's, make all possible attempts to find a common ground and settle your dispute before resorting to lawyers and courts.

Better yet, follow the advice in this book on how to choose a contractor; you'll enjoy a relationship that is both satisfying and productive.

Appendix A
State-by-State Information

In addition to basic contractor requirements, if a statewide contractor's licensing program is in effect, a brief description of the requirements are given, as well as the name, address, and telephone number of the regulatory agency. States having no licensing laws are so noted.

Alabama

Requires statewide licensing for commercial contractors. No testing, bond, or insurance required.

Contact: State Licensing Board for General Contractors
125 S. Ripley Street
Montgomery, AL 36130
(205) 242-2839

Alaska

Requires statewide licensing of all contractors. No test required, although pending legislation will soon require testing of general contractors doing residential work. Contractor must be bonded and insured.

Contact: Department of Commerce and Economic Development
P.O. Box D
Juneau, AK 99811-2534
(907) 465-2535

Arizona

Requires statewide licensing of all contractors. Requires four years' experience and completion of licensing test. Contractor must be bonded and insured.

Contact: Arizona Registrar of Contractors
800 W. Washington, 6th Floor
Phoenix, AZ 85007
(602) 542-1525

Arkansas

Requires license for nonresidential contractors. Applicants must show experience, pass a licensing test, and post a bond. Insurance is optional. Residential construction is not state regulated.

Contact: Contractors Licensing Board
621 E. Capitol
Little Rock, AR 72202
(501) 372-4661

California

Requires statewide licensing for all contractors. Requires four years' verified experience and completion of licensing test. Requires contractor have both bond and insurance.

Contact: Contractor's State Licensing Board
9835 Goethe Road
Sacramento, CA 95827
(916) 366-5247

Colorado

No statewide licensing of building contractors.

Contact: Building department in individual city or county where construction is to be performed.

Connecticut

Requires statewide registration of home-improvement contractors. No experience or testing required. Contractors must carry worker's compensation insurance but no liability insurance required. No bonding required; contractors pay annually into a home-improvement warranty fund.

Contact: Department of Consumer Protection
Occupational Licensing Division
165 Capital Avenue
Hartford, CT 06106
(203) 566-2822

Delaware

Requires statewide licensing for all contractors. Some licenses also issued by individual cities and counties. No experience or testing required. Contractors must be bonded and insured.

Contact: Division of Revenue
P.O. Box 2340
Wilmington, DE 19899
(302) 571-5800

Florida

Requires statewide licensing of all contractors. Requires experience and completion of test. Requires contractor have bond and insurance.

Contact: Construction Industry Licensing Board
P.O. Box 2
Jacksonville, FL 32202
(904) 359-6310

Georgia

Requires statewide licensing of plumbing, electrical, heating and air-conditioning, alarm, and utility contractors. Experience and testing required. No insurance or bond required.

Contact: Georgia Construction Industry Licensing Board
166 Pryor Street, SW
Atlanta, GA 30303
(404) 656-3939

City or county where construction work is to be performed.

Hawaii

Requires statewide licensing for all types of contractors. Requires successful completion of licensing test. Bonding required for some classifications of contractors. Requires workmen's compensation insurance; other insurance optional.

Contact: Department of Commerce and Consumer Affairs
Board of Contractors
1010 Richards Street
Honolulu, HI 96813
(808) 549-4100

Idaho

Requires statewide license for public works contractors only. Some cities also license residential contractors. Requires open-book license test; no bond or insurance.

Contact: Contractors Licensing Board
500 S. 10th Street
Boise, ID 83720
(208) 334-2966

Individual city or county where construction work is to be done.

Illinois

No statewide licensing of building contractors.

Contact: Building department in individual city or county where construction is to be performed.

Indiana

No statewide licensing of building contractors.

Contact: Building department in individual city or county where construction is to be performed.

Iowa

Requires statewide licensing of all contractors. No testing or experience required. Contractor must be insured and bonded.

Contact: Division of Labor
1000 E. Grand Avenue
Des Moines, IA 50319-0209
(515) 281-3636

Kansas

No statewide licensing of building contractors.

Contact: Building department in individual city or county where construction is to be performed.

Kentucky

Requires statewide licensing for plumbing, sprinkler, and boiler contractors only. Requires completion of licensing test. Insurance required; no bond.

Contact: Department of Housing, Buildings, and Construction
Division of Building Codes Enforcement
1047 U.S. 127 South, The 127 Building
Frankfort, KY 40601
(502) 296-3168

Louisiana

Requires contractors be licensed for construction projects valued at $50,000 or more. Requires asbestos removal and hazardous waste contractors be licensed. Licensing test is required for all of these categories. No experience required. Bonding and insurance are optional.

Contact: Department of Commerce
              Contractor's Licensing Board
              7434 Perkins Road
              Baton Rouge, LA 70808
              (504) 765-2301

Maine

Requires statewide licensing for plumbers and electricians.

Contact: Professional and Financial Regulation Department
              License and Enforcement Division
              State House
              Augusta, ME 04330
              (207) 289-3671

Maryland

Requires statewide licensing of home-improvement contractors. Licensing test required; workmen's compensation insurance required; bonding and other insurance optional.

Contact: Department of Licensing and Regulation
              Home Improvement Commission
              501 St. Paul Place
              Baltimore, MD 21202
              (301) 333-8120

Also requires that contractors engaged in new residential construction be licensed, which is essentially a revenue license. No test required. Requires worker's compensation insurance; bonding optional.

Contact: Department of Revenue
              301 W. Preston Street
              Baltimore, MD 21201
              (301) 225-1550

Massachusetts

Requires statewide licensing for all contractors. Contractor must be at least eighteen years old, show three years' verified experience, and pass a licensing test. Bond and insurance are optional.

Contact: Commonwealth of Massachusetts
Registration Division
1 Ashburn Place
Boston, MA 02108
(617) 727-3200

Michigan

Requires statewide licensing for residential and commercial construction and remodeling. Requires licensing test. No insurance or bond required.

Contact: Department of Licensing and Registration
Builders Unit
P.O. Box 30245, 611 W. Ottowa
Lansing, MI 48909
(517) 373-0678

Minnesota

No statewide licensing of building contractors.

Contact: Building department in individual city or county where construction is to be performed.

Mississippi

Requires statewide licensing for contractors doing public work valued at over $100,000 or private work valued at over $50,000. Requires licensing test. No bond or insurance required.

Contact: Mississippi State Contractor's Board
2001 Airport Road, Suite 101
Jackson, MS 39201
(601) 354-6161

Missouri

No statewide licensing of building contractors.

Contact: Building department in individual city or county where construction is to be performed.

Montana

Requires licensing for electrical, plumbing, and public works contractors only. No test, bond, or insurance required.

    Contact: Montana Department of Commerce
                Building Codes Bureau
                1218 E. 6th Avenue
                Helena, MT 59620
                (406) 444-3933

Nebraska

No statewide licensing of building contractors.

    Contact: Building department in individual city or county where construction is to be performed.

Nevada

Requires statewide licensing for all contractors. Requires four years' verified experience and completion of licensing test. Requires contractor have both bond and insurance.

    Contact: State Contractor's Board
                70 Linden Street
                Reno, NV 89502
                (702) 789-0141

                State Contractor's Board
                1800 Industrial Road
                Las Vegas, NV 89158
                (702) 486-3500

New Hampshire

No statewide licensing of building contractors.

    Contact: Building department in individual city or county where construction is to be performed.

New Jersey

Requires statewide registration for new construction contractors. Requires insurance and bond covering a ten-year limited warranty on the home. No requirements for remodeling contractors.

Contact: Bureau of Home Owner Protection
New Jersey Department of Community Affairs
3131 Princeton Pike
Lawrenceville, NJ 08648
(609) 292-5340

New Mexico

Requires statewide licensing of all contractors. Requires two to four years' experience and completion of a licensing test. Contractors required by state labor department to be insured. Bond is one option of proof of financial responsibility.

Contact: Regulation and Licensing Department
Construction Industries Division
725 St. Michaels Drive
Santa Fe, NM 87501
(505) 827-7030 or 827-7059

Regulation and Licensing Department
Construction Industries Division
P.O. Box 25101
Santa Fe, NM 87504

New York

No statewide licensing of building contractors.
Contact: Building department in individual city or county where construction is to be performed.

North Carolina

Requires statewide licensing for contractors. No experience required, but requires licensing test. No bond or insurance required.

Contact: North Carolina Licensing Board for General Contractors
P.O. Box 17187
Raleigh, NC 27619
(919) 781-8771

North Dakota

Requires statewide licensing of all contractors. No experience or testing required. Requires workmen's compensation and liability insurance; contractor must be bonded.

Contact: Secretary of State
Capitol Building
600 E. Boulevard Avenue
Bismarck, ND 58505-0500
(701) 224-3665

Ohio

No statewide licensing of building contractors.

Contact: Building department in individual city or county where construction is to be performed.

Oklahoma

Requires statewide licensing for plumbing, electrical, heating and air-conditioning, and alarm contractors only. Requires completion of a test. Requires insurance and bond.

Contact: Oklahoma Department of Health
1000 NE 10th, Room 807
Oklahoma City, OK 73152
(405) 521-3279

Oregon

Requires statewide licensing of all contractors. No experience or testing required. Requires workmen's compensation and liability insurance; contractor must be bonded.

Contact: Construction Contractor's Board
             700 Summer Street, NE
             Salem, OR 97310
             (503) 378-4621

Pennsylvania

No statewide licensing of building contractors.
Contact: Building department in individual city or county where construction is to be performed.

Rhode Island

Requires statewide registration of contractors. No experience or test required. Contractor must carry insurance; no bond.
Contact: Department of Labor
             Contractor Registration
             220 Elmwood Avenue
             Providence, RI 02907
             (401) 457-1860

South Carolina

Requires statewide licensing of commercial, public works, electrical, plumbing, and heating contractors only. Experience and testing required; contractor must be bonded and insured.
Contact: South Carolina Board of Contractors
             P.O. Box 5737
             Columbia, SC 29250
             (803) 734-8954

South Dakota

Requires statewide licensing for plumbing and electrical contractors. Must verify experience and pass licensing test.

Contact: Department of Commerce and Regulation
Professional and Occupational Licensing
Capitol Building
Pierre, SD 57501
(605) 773-3177

Tennessee

Requires statewide license for contractors doing work that is valued at $25,000 or more. Requires completion of test. No bond or insurance required.

Contact: Contractor's Licensing Board
500 James Robertson Parkway, Suite 110
Nashville, TN 37243-1150
(615) 741-2121

Texas

No statewide licensing of building contractors.

Contact: Building department in individual city or county where construction is to be performed.

Utah

Requires statewide licensing of all contractors. Requires four years' verified experience and completion of licensing test. Contractors must be insured; no bond required.

Contact: Division of Occupational and Professional Licensing
P.O. Box 45802
Salt Lake City, UT 84145
(801) 530-6514

Vermont

No statewide licensing of building contractors.

Contact: Building department in individual city or county where construction is to be performed.

Virginia

Requires statewide licensing for all contractors. Requires testing for plumbing, electrical, and heating and air-conditioning contractors; all other contractors will be tested as of January 1, 1991. Requires insurance; may require bond at local level.

Contact: Contractor's Board
            3600 W. Broad Street
            Richmond, VA 23230
            (804) 367-8511

Washington

Requires statewide licensing of all contractors. No licensing test required. Contractors must be bonded and insured.

Contact: Department of Labor and Industries
            Contractor's Registration
            P.O. Box 9689, 805 Plum Street
            Olympia, WA 98504
            (206) 568-8046

West Virginia

No statewide licensing of building contractors.
Contact: Building department in individual city or county where construction is to be performed.

Wisconsin

No statewide licensing of building contractors.
Contact: Building department in individual city or county where construction is to be performed.

Wyoming

No statewide licensing of building contractors.
Contact: Building department in individual city or county where construction is to be performed.

Appendix B
For More Information

The following organizations may be able to provide additional information about specific areas of contractor relations and dispute resolution. Many offer catalogs of publications, as well as brochures of general information.

Dispute Resolution

American Arbitration Association (AAA)
140 W. 51st Street
New York, NY 10020
(212) 484-1400

American Bar Association
Standing Committee on Dispute Resolution
1800 M Street NW, Suite 200
Washington, DC 20036
(202) 331-2258

National Academy of Conciliators
7315 Wisconsin Avenue, Suite 1255N
Bethesda, MD 20814
(301) 907-7000

National Institute for Dispute Resolution
1901 L Street NW, Suite 600
Washington, DC 20036
(202) 446-4764

Consumer Information

Consumer Information Center
18 F Street NW, Room G-142
Washington, DC 20405
(202) 566-1794

Council of Better Business Bureaus
4200 Wilson Boulevard, Suite 800
Arlington, VA 22203
(703) 276-0100

Contractor Associations

International Remodeling Contractors Association (IRMA)
P.O. Box 17063
West Hartford, CT 06117
(203) 233-7442

National Association of Home Builders (NAHB)
15th and M Streets NW
Washington, DC 20005
(202) 822-0200

National Kitchen and Bath Association (NKBA)
124 Main Street
Hackettstown, NJ 07840
(201) 852-0033

Architects and Designers

American Institute of Architects (AIA)
1735 New York Avenue NW
Washington, DC 20006
(202) 626-7300

American Society of Interior Designers (ASID)
1430 Broadway
New York, NY 10018
(212) 944-9200

Appendix C
Sample Forms

1. A simple contractor bidding sheet, typically used for remodeling and repair estimates.

JOB _____ BID _____ DATE _____
_____ TO _____ PH-H _____
_____ _____ PH-W _____

DESCRIPTION _____

S	ITEM	NOTES	OS	COST
	PLANS			
	BUILDING PERMIT			
	DEMOLITION			
	DEBRIS REMOVAL			
	EXCAVATION			
	FOUNDATION			
	FLATWORK			
	STRUCTURAL-MATERIAL			
	-LABOR			
	FASTENER/HARDWARE			
	ROOF TRUSSES			
	SIDING			
	ROOFING			
	GUTTERS/SHEET METAL			
	WINDOWS			
	GLASS DOORS			
	DOORS			
	MILLWORK			
	ELECTRICAL-RGH			
	-TOP			
	PLUMBING-RGH			
	-TOP			
	MIRROR/TOWEL BAR			
	HVAC			
	INSULATION			
	DRYWALL/PLASTER			
	PAINT-INTERIOR			
	-EXTERIOR			
	OTHER WALLCOVER			
	CABINETS			
	COUNTERS			
	APPLIANCES			
	UNDERLAYMENT			
	CARPET/LINOLEUM			
	CERAMIC TILE/HARDWOOD			
	MASONRY			
	RENTAL EQUIPMENT			
	CRANE SERVICE			

SUBTOTAL	
CONTINGENCY (%)	
PROFIT/OVERHEAD (%)	
TOTAL	

2. A more complex bidding sheet of the type used by many contractors for bidding new homes.

DESCRIPTION OF MATERIALS AND ITEMIZATION OF COSTS

BUILDER OR CONTRACTOR _____ PHONE Home _____

Bus. _____

OWNER _____ PHONE Home _____

Bus. _____

ADDRESS OF PROPERTY _____ Bldr. Lic. # _____

LOT: _____ BLOCK: _____ ADDITION: _____ COUNTY: _____

Lot Set Backs: _____ Front: _____Ft.   Sides: _____ _____Ft.   Back _____ _____Ft.
Zoning Classification: _____ _____

MATERIALS	COST

1. PLANS, PERMITS, ARCHITECT'S FEES .... $ _____
2. BONDS, INSURANCE .... $ _____
3. EXCAVATION, BACKFILL, FINAL GRADING .... $ _____
4. FOUNDATION ....MATERIAL: $ _____
                                                                          LABOR: $ _____

Bearing soil type _____
Foundation material   ☐ Concrete   ☐ Concrete Block
                       ☐ Piles   ☐ Piers
                       ☐ Slab   Depth _____
Moisture Barrier—describe _____
Drainage System _____

5. FRAMING, SHEATHING ....MATERIAL: $ _____
                                                                          LABOR: $ _____

☐ Single Construction   ☐ Double Construction
Posts   ____x____ Grade _____Material _____
Girders   ____x____ Grade _____Material _____
Joists   ____x____ Grade _____Material _____
Subfloor   ____x____ Grade _____Material _____
Studs   ____x____ Grade _____Material _____
Rafters   ____x____ Grade _____Material _____
Trusses _____
Sheathing: Roof _____ Walls _____

6. ROOFING .... $ _____

Waterproofing _____
☐ Cedar shingles   ☐ Cedar shakes   ☐ Composition
☐ Tile   ☐ Other _____

7. SIDING ....MATERIAL: $ _____
                                                                          LABOR: $ _____

☐ Brick   ☐ Aluminum   Thickness _____
☐ Stone   ☐ Plywood   Explain _____
☐ T1-11   ☐ Wood   Type _____
☐ Other _____

8. WINDOWS, SASH .... $ _____

Make _____   Type _____
☐ Double Glazed   ☐ Screens
Describe _____   ☐ Storm Windows

9. BRICKWORK .... $ _____

Firebox: ☐ Brick   ☐ Metal   Number of Fireplaces (_____)
          ☐ Free Standing   ☐ Wood Stove(s)   # _____

10. SHEET METAL WORK .... $ _____

Gutters/Downspouts—describe _____
Connected to _____

11. EXTERIOR/INTERIOR PAINTING AND DECORATING .... $ _____

Ext. no. of coats _____   Materials _____
Int. no. of coats _____   Materials _____
Paneling—describe _____
Wallpaper—describe _____

12. PLUMBING AND PLUMBING FIXTURES . . . . $ _____

   (Please indicate if Fiberglass, Stainless Steel, or Cast Iron under "Type")
   Fixtures:
   Main Bath _____ Type _____ Mfgr. _____
   2nd Bath _____ Type _____ Mfgr. _____
   3rd Bath _____ Type _____ Mfgr. _____
   Kitchen _____ Type _____ Mfgr. _____
   Utility _____ Type _____ Mfgr. _____
   Other: (i.e., Sauna, Jacuzzi, Hot Tub, etc.) _____

13. WATER . . . . $ _____

   ☐ Public   ☐ Community   ☐ Well-private   ☐ Solar
   Water District _____

14. SEWER . . . . $ _____

   ☐ Sewer hookup   ☐ Septic   ☐ Cesspool

15. ELECTRICAL WIRING AND FIXTURES . . . . $ _____

   Installed by _____
   ☐ Surface   ☐ Underground
   No. of outlets (110) _____   No. of outlets (220) _____
   Fixture allowance _____

16. INSULATION . . . . $ _____

   Ceiling: Material _____ R Factor _____
   Walls: Material _____ R Factor _____
   Floor (over crawl space)   ☐ Under Subfloor
   ☐ Foundation Perimeter
         Material _____ R Factor _____
   Basement walls: Material _____ R Factor _____
   Vapor barrier: Material _____ R Factor _____
   Weatherstripping—describe _____

17. HEATING, COOLING . . . . $ _____

   Type: Fuel:
   ☐ Forced Air   ☐ Electric
   ☐ Baseboard   ☐ Gas
   ☐ Ceiling cable   ☐ Oil
   ☐ Radiator   ☐ Hot water or water
   ☐ Wall unit   ☐ Wood
   ☐ Heat pump   ☐ Coal
   ☐ Other _____   ☐ Solar—please enclose specs.

                                  BTU's _____

   Air Conditioner: ☐ Central   ☐ Wall   ☐ Window
   Ductwork insulation _____   Thickness _____

18. INTERIOR WALLS/DRYWALLS . . . . $ _____

   Describe finish: walls _____
   Describe finish: ceilings _____
   Other walls—describe _____

19. DOORS, TRIM, FINISH . . . . $ _____

   Interior doors: Type _____
   Exterior doors: Type—Front _____   Rear _____
   Trim: Wood _____   Vinyl _____   Sheetrock _____
   Trim finish: Material _____   No. of Coats _____

20. GARAGE MATERIALS . . . . $ _____

   Garage door—describe _____
   Automatic opener—describe _____
   Garage interior, full-sheetrocked _____
   Garage interior, firewall only _____

21. CABINETS . . . . $ _____

   Mfgr. _____
   Materials: ☐ Plywood   ☐ Particle board
   Veneer—describe _____
   Finish: Materials _____   No. of coats _____

125

22. VINYL, FORMICA, TILEWORK .... $ _____

      Countertops: Formica _____   Other _____
      Flooring: Kitchen _____
                Baths _____
      Tilework: Kitchen _____
                Baths _____
      Entry _____

23. APPLIANCES .... $ _____

      Range: □ Free standing   □ Drop-in
                        Make _____ Model # _____
      Oven: □ Free Standing   □ Drop-in   □ Built-in
                        Make _____ Model # _____
      Dishwasher .... Make _____ Model # _____
      Disposal .... Make _____ Model # _____
      Hood/Fan .... Make _____ Model # _____
      Jennair .... Model # _____
      Microwave .... Make _____ Model # _____
      Intercom .... Make _____ Model # _____
      Vacuum .... Make _____ Model # _____
      Trash compactor .... Make _____ Model # _____
      Bath fans .... Make _____ Model # _____
      Smoke Detector .... Make _____ Model # _____
      Security System .... Make _____ Model # _____
      Other .... Make _____ Model # _____

24. UNDERLAYMENT .... $ _____

      □ Particle board   □ Plywood   Thickness _____

25. FLOORING .... $ _____

      □ Carpet—Located _____
      □ Hardwood—Located _____
      Other—Located _____

26. PORCHES, PATIOS, ETC. .... $ _____

      □ Walk
      □ Porch
      □ Deck Size _____
      □ Patio Size _____
      □ Other _____
      Please check (✓) box where appropriate and **give size when requested to receive value.

27. DRIVEWAY—SQUARE FEET _____ .... $ _____

      □ Gravel   □ Asphalt   □ Concrete
      □ Other _____

28. LANDSCAPING .... $ _____

      □ Barkdust   □ Shrubs   □ Lawn   □ Retaining Walls

29. MISCELLANEOUS EXTRAS .... $ _____

      □ Rails   □ Wrought Iron   □ Sprinkler System
      □ Planters   □ Other _____

30. CLEANUP .... $ _____

31. FINANCING .... $ _____

32. REAL ESTATE COMMISSIONS .... $ _____

33. PROFIT AND OVERHEAD .... $ _____

34. LOT (Date acquired _____   Purchase price _____)—VALUE: $ _____

                                            TOTAL COSTS: $ _____

      FLOOR PLAN: Square feet — Basement Level _____
                            Main Level _____
                            Upper Level _____
                            Loft _____

INDICATE ROOMS TO BE **FINISHED ONLY**

BASEMENT LEVEL	MAIN LEVEL	UPPER LEVEL	LOFT
☐ Bath (1)	☐ Living room	☐ Bath (1)	☐ Bath (1)
☐ Bath (2)	☐ Dining room	☐ Bath (2)	☐ Bath (2)
☐ Bedroom (1)	☐ Bath (1)	☐ Bedroom (1)	☐ Bedroom (1)
☐ Bedroom (2)	☐ Bath (2)	☐ Bedroom (2)	☐ Bedroom (2)
☐ Bedroom (3)	☐ Bath (3)	☐ Bedroom (3)	☐ Study
☐ Hall	☐ Bedroom (1)	☐ Bedroom (4)	☐ Den/Family Room
☐ Den/Family Room	☐ Bedroom (2)	☐ Utility	☐ Other _____
☐ Garage	☐ Bedroom (3)	☐ Other _____	
☐ Utility	☐ Bedroom (4)		
☐ Other _____	☐ Kitchen		
	☐ Den/Family Room		
	☐ Garage		
	☐ Utility		
	☐ Other _____		

If house is to be built under contract, describe any work to be done by owner:

_____

_____

How is cost of this work to be handled? _____

_____

 I do state that I have prepared the foregoing description of materials with applicable costs to be used in construction. I certify that such materials by grade or better and in the quantities and dollar equivalents stipulated will be used in such construction, and that all work shall be performed in a good and workmanlike manner meeting in every respect all applicable code requirements. I agree to call for and obtain all code or other applicable inspections as required. Further, I agree to immediately notify and obtain the approval of _____ of any structural, design, layout, or major material or cost changes from the foregoing description and attach plans which are contemplated. This certificate is given to induce _____ to grant a mortgage loan, the proceeds of which will be used to pay construction costs.

DATED _____ 19_____

_____
Builder/Contractor

AGREED TO AND ACCEPTED BY:

_____
Owner

127

3. A simplified stock proposal and contract form.

PROPOSAL and CONTRACT

Date _____, 19 _____

TO _____

_____

Dear Sir:

_____ propose to furnish all materials and perform all labor necessary to complete the following:

_____

_____

_____

_____

_____

_____

_____

_____

_____

All of the above work to be completed in a substantial and workmanlike manner according to standard practices for the sum of _____ Dollars ($ _____)

Payments to be made _____

_____

_____ as the work progresses to the value of _____ percent (_____ %) of all work completed. The entire amount of contract to be paid within _____ days after completion.

Any alteration or deviation from the above specifications involving extra cost of materials or labor will be executed only upon written orders for same, and will become an extra charge over the sum mentioned in this contract. All agreements must be made in writing.

Respectfully submitted,

By _____

ACCEPTANCE

You are hereby authorized to furnish all materials and labor required to complete the work mentioned in the above proposal, for which _____ agree to pay the amount mentioned in said proposal, and according to the terms thereof.

ACCEPTED _____

_____

Date _____, 19 _____

4. A typical attorney-prepared construction contract.

S SPECTRUM CONSTRUCTION

Full Service Builders
P.O. Box 968
Bend, Oregon 97709

Contractors Lic. #58997
(503) 385-5310
(503) 382-1622

PROPOSAL AND CONTRACT

TO: _____ AT: _____

Dear _____

    SPECTRUM CONSTRUCTION proposes to furnish all materials and perform all labor necessary to complete the following: _____

_____

(All further materials and labor not specified above shall be further set forth in attachments to this contract)

All of the above work (and any work set forth in attachments, if applicable), shall be completed in a substantial and workmanlike manner according to standard practices for the total sum of $ _____
(_____ Dollars)

Payment shall be made according to the following schedule:
      Amount   Due

_____   _____

_____   _____

_____   _____

    The approximate date when the work herein described shall commence shall be _____
and the approximate date the work shall be substantially completed shall be _____ .
    Final payment, constituting the entire unpaid balance of this contract and including the amount of all change orders, shall be made not later than 15 days after completion. Completion shall be considered to occur upon satisfactory final inspection by a building inspector; occupancy of the improvement by the owner or abandonment as defined by ORS 87.045; or the posting of a completion notice, whichever occurs first.

    Any alteration or deviation from the specifications set forth in this Proposal and Contract involving increased or reduced cost of materials or labor will only be executed upon written orders for same, and will become an additional charge or a credit to the sums set forth in this agreement. Additional labor and materials required by the excavation and removal of rock, or by changes required by building department officials, will be specified in writing and will become an additional charge over and above this contract amount. SPECTRUM CONSTRUCTION assumes no liability and offers no warranty on labor or materials, new or used, which are supplied by the owner.

                                            Respectfully submitted,

                                            _____

Date _____   By _____
                                             Co-Owner, Spectrum Construction

ACCEPTANCE

    SPECTRUM CONSTRUCTION is hereby authorized and directed to furnish all materials and labor required to complete the work as set forth in the Proposal hereinabove set forth (and in Attachments, if applicable), for which the undersigned jointly and severally agree to pay the amount mentioned in said Proposal and according to the terms thereof.

    The undersigned further agrees that in the event the undersigned defaults or fails to make payment(s) as herein agreed, the undersigned will pay all reasonable attorney fees and costs, including collection costs, necessitated by said default to enforce this contract. Default or non-payment is defined as failure to make payments to SPECTRUM CONSTRUCTION or their order within 30 days of the due date. Further, the undersigned agrees that for any balance due under this agreement outstanding for a period of more than 30 days after due date, a late charge of 5% per month of the balance due shall be charged and added to any and all outstanding amounts.

    The undersigned further represents and acknowledges they have fully read and understand the terms and conditions of this Proposal and Contract and FULLY ACCEPT EACH AND EVERY TERM AND CONDITION herein. No other promises or acts have been contracted for other than as set forth in this Proposal and Contract.

Dated: _____   _____

                                               _____

SPECTRUM CONSTRUCTION has provided an "Information Notice To Owners About Construction Liens." _____
                                                                  Initial

5. A typical claim of lien document, used by a contractor or material supplier to file a lien against a property for nonpayment.

Claim of Lien

In   Court,

County of

State of

vs.

Dated

FLA. 1980 MECHANICS'
LIEN LAW, FS 713.08

RAMCO FORM 120½

SPACE ABOVE THIS LINE FOR PROCESSING DATA — — SPACE ABOVE THIS LINE FOR RECORDING DATA —

Claim of Lien

State of Florida
County of

Before me, the undersigned Notary Public, personally appeared ....

who was duly sworn and says that he is (the lienor herein) (the agent of the lienor herein) ....
(DELETE ONE)

....
(Lienor's Name)

whose address is .... ,
(Lienor's Address)

and that in accordance with a contract with .... ,

....

lienor furnished labor, services or materials consisting of: (Describe specially fabricated materials separately)

on the following described real property in ....County, Florida:
(Describe real property sufficiently for identification, including street and number, if known)

owned by ....

of a total value of ....dollars ($ ....)

of which there remains unpaid $ .... and furnished the first of the items on

.............................., 19 .... and the last of the items on ...., and (if the lien

is claimed by one not in privity with the owner) that the lienor served his notice to owner on ....

.........................., 19 ...., by .... ,
(Method of Service)

and, (if required) that the lienor served copies of the notice on the contractor on

by ...., and on the subcontractor, ...
(Method of Service)

on ...., 19 ...., by ....
(Method of Service)

....
Lienor

By ....
Agent

Sworn to and subscribed before me this..........................day of .... 19 ....

....
Notary Public

6. A preliminary lien notice. Note that it specifically says it is not a lien but rather a notification of the right to lien.

NOTICE OF THE RIGHT TO LIEN
WARNING: READ THIS NOTICE. PROTECT YOURSELF FROM PAYING
ANY CONTRACTOR OR SUPPLIER FOR THE SAME SERVICE.

Date of Mailing: _____

TO: _____
_____
_____

THIS IS TO INFORM YOU that _____
_____

has begun to provide (description of materials) _____
_____
_____
_____
_____

ordered by _____
_____

for improvements to property you own. The property is located at _____
_____

A lien may be claimed for all materials, labor, and services furnished after a date that is eight days not including Saturdays, Sundays, and other holidays as defined in ORS 187.010 before this notice was mailed to you.

Even if you or your mortgage lender have made full payment to the contractor who ordered these materials or services, your property may still be subject to a lien unless the supplier providing this notice is paid.

THIS IS NOT A LIEN. It is a notice sent to you for your protection in compliance with the construction lien laws of the State of Oregon.

This notice has been sent to you by

Name: _____
Address: _____
_____
Telephone: _____

IF YOU HAVE ANY QUESTIONS ABOUT THIS NOTICE, FEEL FREE TO CALL US.

SEE OVER FOR IMPORTANT INFORMATION

Under Oregon's laws, those who work on your property or provide materials and are not paid have a right to enforce their claim for payment against your property. This claim is known as a construction lien.

If your contractor fails to pay subcontractors, material suppliers, or laborers or neglects to make other legally required payments, the people who are owed money can look to your property for payment, <u>even if you have paid your contractor in full.</u>

The law states that all people hired by a contractor to provide you with materials, labor, or services must give you a notice of the right to lien to let you know what they have provided.

WAYS TO PROTECT YOURSELF ARE

- RECOGNIZE that this notice of delivery of materials, labor, or services may result in a lien against your property unless all those supplying a notice of the right to lien have been paid.

- LEARN more about the lien laws and the meaning of this notice by contacting the Builders Board, an attorney, or the firm sending this notice.

- ASK for a statement of the labor or materials provided to your property from each party that sends you a notice of the right to lien.

- WHEN PAYING your contractor for materials, labor, or services, you may make checks payable <u>jointly</u> to the contractor and the firm furnishing materials, labor, or services for which you have received a notice of the right to lien.

- OR use one of the methods suggested by the "Information Notice to Owners." If you have not received such a notice, contact the Builders Board.

- GET EVIDENCE that all firms from whom you have received a notice of the right to lien have been paid or have <u>waived</u> the right to claim a lien against your property.

- CONSULT an attorney, a professional escrow company, or your mortgage lender.

7. Another type of preliminary lien notice, this one for the delivery of materials to a job.

NOTICE TO MORTGAGEE OR BENEFICIARY UNDER TRUST DEED OF DELIVERY OF MATERIAL AND SUPPLIES

TO ....

....

    This is to advise you, the owner of record of a mortgage or a beneficiary in a trust deed, on either the said land or improvements thereon, that the undersigned is delivering materials and supplies upon the order of

....

....

for use in the construction of an improvement located upon the following

described site in .... County, Oregon:

....

....

....

....

also known as .... Oregon.
STREET ADDRESS IF KNOWN

    You are further notified that a lien may be claimed for all such materials and supplies so delivered, after a date that is 8 days, not including Saturdays, Sundays and other holidays, as defined in ORS 187.010 before this notice is delivered to you in person or mailed to you by registered or certified mail and that payment by the owner or lender to the contractor does not remove the right of the undersigned furnishing such materials or supplies to claim a lien against the above described property unless the undersigned is in fact paid. No further notice to you of this or any subsequent delivery is necessary.

....
NAME

....
ADDRESS

Delivered to you by registered or certified mail	When delivered in person:
at ...., Oregon	Receipt of above notice is acknowledged.
on ...., 19........	Dated: ...., 19........
....	....
....	....
	MORTGAGEE—BENEFICIARY UNDER TRUST DEED

ORS 87.025(3) provides: (3) No lien for materials or supplies shall have priority over any recorded mortgage *or trust deed* on either the land or *improvement* unless the person furnishing *the* material or supplies, not later than 8 days, not including Saturdays, Sundays or other holidays as defined in ORS 187.010, after the date of delivery of material or supplies for which a lien may be claimed, delivers to the *mortgagee*, a notice in the form required by ORS 87.023 or substantially the same information as said Form.

FORM No. 1160A Ⓒ 1983
Stevens-Ness Law Pub. Co.
Portland, OR 97204 OA

NOTE: THIS FORM TO BE USED ONLY FOR CONSTRUCTION COMMENCED AFTER OCTOBER 14, 1983.

8. To protect yourself against liens.

WAYS TO PROTECT YOURSELF

- If you are dealing with a lending institution, ask your loan officer what precautions the institution takes when disbursing mortgage money to your contractor to verify that subcontractors and material suppliers are being paid.

- If you are paying your contractor directly, request a current statement of labor or materials provided to your property from each party that has sent you a Notice of the Right to Lien. You should make this request in writing and send it by certified mail. The party sending this Notice is required by law to respond to your request within 15 days from the date your letter is received.

- Make your check payable jointly. Name the contractor and the subcontractor or supplier as payees.

- Ask your contractor for a lien waiver from each party who has sent you a Notice of the Right to Lien.

- Consider using the services of an escrow agent to protect your interests. Find out whether your escrow agent will protect you against liens when disbursing payments. If you are interested in this alternative, consult your attorney.

When in doubt or if you need more details, consult an attorney. When and how to pay your contractor is a decision to which you should give serious thought.

By signing this notice you are indicating that you have received this notice, have read it and understand it. Your signature does not, in any way, give your contractor or those who provide material, labor or services any additional rights to place a lien on your property.

Job Site Address: _____

This notice was furnished by: This notice was received by:

_____   _____
Contractor   Property Owner

_____ _____   _____
Builders Board   Date   Date
Registration Number

If you find yourself in a "pay twice" situation, help may be available to you through the Builders Board. You may be able to file a claim with that agency.

For more details about the assistance available through the Builders Board, you may write to:

Builders Board
Department of Commerce
403 Labor and Industries Building
Salem, Oregon 97310
378-4621

The material in this notice is not intended to be a complete analysis of the law (ORS Chapters 87 and 701). For more detailed information, contact your attorney.

---

9. A typical information notice of the type required by some states to inform consumers about construction liens.

Department of Commerce
BUILDERS BOARD

403 LABOR & INDUSTRIES BUILDING, SALEM, OREGON 97310   PHONE 378-4621

VICTOR ATIYEH
GOVERNOR

> OREGON LAW REQUIRES YOUR CONTRACTOR TO GIVE YOU THIS NOTICE if your contract price exceeds $1,000. The purpose of this notice is to explain the basics of the construction lien law and to help you protect yourself. This notice is not a reflection upon the abilities or credit of your contractor.

INFORMATION NOTICE TO OWNERS
ABOUT CONSTRUCTION LIENS

IF YOUR CONTRACTOR FAILS TO PAY SUBCONTRACTORS, MATERIAL SUPPLIERS OR LABORERS OR NEGLECTS TO MAKE OTHER LEGALLY REQUIRED PAYMENTS, THOSE PEOPLE WHO ARE OWED MONEY CAN LOOK TO YOUR PROPERTY FOR PAYMENT, EVEN IF YOU HAVE PAID YOUR CONTRACTOR IN FULL. THIS IS TRUE IF YOU:

> **HAVE HIRED a contractor to build a new home;**
>
> **ARE BUYING a newly-built home;**
>
> **ARE REMODELING or improving your property.**

Under Oregon's laws, those who work on your property or provide materials and are not paid have a right to enforce their claim for payment against your property. This claim is known as a construction lien.

Persons who supply materials or labor ordered by your contractor are permitted by law to file a lien only if they have sent to you a Notice of the Right to Lien.

If you enter into a contract to buy a newly-built home or a partly-built home, you may not receive a Notice of the Right to Lien. Be aware that a lien may be claimed even though you have not received notice. You may want to ask your contractor or title insurance company about an ALTA title insurance policy based upon the receipt of lien waivers.

You have final responsibility for seeing that all bills are paid even if you have paid your contractor in full.

If you receive a Notice of the Right to Lien, take the Notice seriously. Let your contractor know you have received the Notice. Find out what arrangements are being made to pay the sender of the Notice.

136

Appendix D
Product Ratings

Anyone remodeling a home, whether he or she is hiring a contractor or doing it him- or herself, needs accurate, unbiased information about products and materials that will be used. The Ratings from *Consumer Reports* will help you choose the right products for your needs.

The Ratings

Individual brands and models are rated based on the estimated quality of the tested product samples. The Ratings order is derived from laboratory tests, controlled use tests, and/or expert judgments. Products judged high in quality and appreciably superior to other products tested receive a check(ν) rating. If a product is both high in quality and relatively low in price, it is designated A Best Buy. The Ratings offer comparative buying information that greatly increases the likelihood you will receive value for your money.

When using the Ratings, first read the introduction preceding each chart, then the notes and footnotes in order to find out about the features, qualities, or deficiencies shared by products in the test group. You will often find out, in a sentence that starts with "Except as noted," what qualities the rated products have in common.

The first sentence in the introduction to each Ratings chart tells you the basis of the Ratings order. Sometimes groups of products may be listed alphabetically or by price when the quality differences are so small as to be not worth considering. Usually, however, Ratings are "Listed in order of estimated quality." That means Consumers Union's engineers judged the brand (or product type) listed first to be the best, the one listed next to be second best, and so on. Sometimes, rated products are about equal to one another and are therefore listed in a special fashion, perhaps alphabetically and within brackets.

Prices. Where there is a listing of brands and models, there is a notation of the month and year in which the Ratings appeared in *Consumer Reports*. Usually the prices are listed as published at the time of the original report. Whenever possible, it pays to shop around and buy from the dealer or source that offers the best price and also provides satisfactory return and servicing arrangements.

Model Changes. Even though the particular brand and model you select from the Ratings may be out of stock or superseded by a later version, the information given can be of great help in sorting out products and their characteristics.

Ratings
Waterproofing paints

● ◐ ○ ◑ ●
Better ←——————→ Worse

As published in *Consumer Reports,* February 1990.

Listed in order of estimated quality, based on water resistance.

❶ **Brand and model.** Most are available only in white; footnotes list the exceptions. If more than one color was available, we tested white and beige.

❷ **Type.** The oil-based epoxy liquids (**OEL**) have to be mixed with the catalyst provided and applied within thirty minutes or they become too thick to use. The oil-based liquids (**OL**) come ready to use but should only be applied to dry walls because oil does not adhere to water. Cementlike powders (**P**) have to be mixed with water; if you mix more than you can apply in about four hours, the paint hardens in the pail. The powders are best applied to a wet wall. Water-based liquids (**WL**) are the easiest to use; they are ready-mixed and can be brushed onto damp or dry walls. Water-based epoxy (**WEL**), like the other epoxies, must be mixed with a catalyst.

❸ **Price.** The manufacturer's suggested retail price, rounded to the nearest dollar. These paints are rarely on sale. + means shipping is extra.

❹ **Size.** We list only the weight of the powders. It's hard to say how many gallons of paint they can make because the amount of water to be added varies with each brand.

❺ **Cost of sample basement.** Our estimate, rounded to the nearest $25, of what it costs to apply two coats to 1,120

square feet of concrete wall. These paints are generally comparable in price to interior wall paint, but a gallon doesn't go as far since it has to cover a rough, pitted surface.

❻ **Water resistance.** The bars in the graph show how waterproof these paints were. To measure water resistance, we put two coats of paint on concrete blocks, sealed the openings in each block, then suspended a water tank eight feet overhead; tubing from the tank let water flow into the cavities, to simulate the water pressure exerted at the bottom of a typical basement wall. We checked the condition of the blocks periodically, and weighed the water tank to find out how much water was seeping through the paint. Our statisticians analyzed the leakage-rate data to produce the index of water resistance shown here.

❼ **Ease of application.** How easy we found it to apply each paint, using a short, stiff-bristled brush. Waterproofing paints are much harder to work with than wall paints because you must apply them with a stabbing or scrubbing motion to seal all the pin-size holes in concrete blocks.

❽ **Stain removal.** How easily we could remove a grease stain from a painted block, using a soft brush and an all-purpose household cleaner.

❾ **Surface smoothness.** In our judgment, the smoothest felt like standard wall paint; the roughest, like a concrete driveway. Rough walls may not matter if you use your basement only for storage, but you may want a nicer finish if you use the basement as living space.

Brand and model [1]	Type [2]	Price [3]	Size [4]	Cost of sample basement [5]	Water resistance [6] (0–100)	Ease of application [7]	Stain removal [8]	Surface smoothness [9]
✓ Barrier System Cat. No. 502 Epoxy Resin [1][2]	OEL	$41+	5 qt. [3]	$500	~90	○	⊙	⊙
✓ Atlas Epoxybond Epoxy Waterproof Sealant [1]	OEL	54	3 qt. [3]	825	~90	◐	⊙	●
Glidden Spred Waterproof Basement Paint [4]	OL	15	1 gal.	350	~75	○	●	○
Bondex Waterproof Cement Paint [4]	OL	21	1 gal.	475	~60	○	●	○
Tru-Test Supreme Tru-Seal Waterproofing Masonry Paint [4]	OL	18	1 gal.	400	~65	◐	◐	⊙
UGL Drylok Masonry Waterproofer	OL	18	1 gal.	400	~65	◐	◐	◐
Moore's Waterproofing Masonry Paint [5]	OL	18	1 gal.	500	~45	◐	◐	⊙
Thoro Super Thoroseal Redi-Mix Liquid [4]	OL	57	2 gal.	625	~35	◐	◐	●
Thoro Super Thoroseal [4]	P	29	20 lb.	675	~50	◐	●	●
Quikrete Heavy Duty Masonry Coating	P	12	40 lb.	158	~40	●	●	●
Bondex Waterproof Cement Paint	P	19	25 lb.	100	~15	◐	●	○
UGL Drylok Double Duty Sealer [4]	P	26	35 lb.	350	~15	◐	●	●
Quikrete Waterproofing Masonry Coating	P	10	20 lb.	225	~10	◐	⊙	●
Muralo Tite Vinyl Latex Waterproofing Paint	WL	15	1 gal.	350	~8	⊙	⊙	●
Sears Basement Waterproofing Latex Wall Paint (Series 5640)	WL	11+	1 gal.	250	~5	⊙	⊙	●

[1] Available only as clear sealer. Mfr. of **Atlas** says product can be tinted slightly.
[2] Available from Defender Indo, P.O. Box 820, New Rochelle, N.Y. 10801-0820.
[3] Includes catalyst.
[4] Available in white only.
[5] Available in white only. Mfr. says product can be tinted.

Ratings
Interior semigloss paints

Better ⟵————⟶ Worse

These listings are from a May 1989 report.

Listed by types. Within types, listed alphabetically. Brand-performance judgments apply to all colors within a brand. Dashes indicate that a suitable color wasn't available. Prices are the manufacturer's suggested retail price per gallon, rounded to the nearest dollar.

❶ Gloss. Some "semiglosses" were shinier than others. Here we note our observed judgments of gloss. Low (**L**) appeared merely satiny. High (**H**) approached the shininess of a glossy enamel. Medium (**M**) was in between, and is probably the best choice if you're looking for a clear-cut semigloss finish.

❷ Brushing ease. Alkyd paints are stickier and thus not as easy to brush on as the latexes.

❹ Sagging. A judgment of how much a paint may run or "curtain" when applied with a brush.

❺ Spattering. Paints that scored well here resisted the tendency to spin paint mist off a roller.

Brand and model	Price ❶	Gloss ❶	Brushing ❷	Leveling ❸	Sagging ❹	Spattering ❺	Scrubbing ❻	Water ❼	Blocking ❽	White Color name	One-coat hiding ❾	Two-coat hiding ❾	Fading ❿	Gold Color name	One-coat hiding ❾	Two-coat hiding ❾	Fading ❿
Alkyd (oil-based)																	
Benjamin Moore Satin Impervo Series 235	$25	L	○	◐	○	◐	◐	●	●	White	1	4	○	Golden Glow	3	6	◐
Devoe Velour Series 26XX	27	M	◐	○	○	◐	◐	●	●	White	1	4	○	Swirl	5	6	◐
Dutch Boy Dirt Fighter Series 555XX	18	M	○	○	○	◐	◐	●	●	Diamond White	2	5	◐	Champagne	3	6	◐
Glidden Spred Lustre Series 4600	27	H	○	○	○	◐	●	●	●	White	1	3	◐	Buttered Rum	2	6	◐
Glidden Spred Ultra Series 4200	30	L	○	◐	○	◐	●	●	●	Bright White	2	5	○	Buttered Rum	2	6	◐
Pittsburgh Wallhide Series 27	26	H	○	◐	○	●	○	●	●	White	1	4	◐	Cactus Flower	2	3	◐
Pratt & Lambert Pro-Hide Plus E3800	21	M	○	◐	○	◐	●	●	●	White	2	5	○	Wind Song	5	6	◐
✓ Pratt & Lambert Cellu-Tone Series C30572	30	L	○	◐	◐	◐	●	●	●	One Coat White	2	6	○	Wind Song	5	6	◐
Sherwin-Williams Classic 99	26	L	○	○	◐	◐	●	●	●	Pure White	2	5	◐	Caramel Corn	3	6	●
Tru-Test Supreme Satin W-Line	22	L	○	○	◐	◐	●	●	●	White	2	5	◐	Sunny Mesa	2	6	○
Valspar Semi-Gloss Enamel Series 614	25	M	○	◐	○	◐	◐	●	●	Super White	2	6	●	Javelin	4	6	●
Latex (water-based)																	
Benjamin Moore Regal Aquaglo Series 333	23	L	●	◐	○	●	◐	○	●	Non-Yellowing White	2	5	●	Golden Glow	3	6	○
Devoe Wonder-Tones Interior Series 38XXN	26	H	●	◐	◐	◐	●	○	◐	White	2	4	◐	Swirl	2	6	○
Dutch Boy Super Kem-Tone	18	L	●	○	◐	●	●	◐	●	White	1	2	●	Wheat Grain	4	6	●
Dutch Boy Dirt Fighter Series 73XX	18	L	●	◐	◐	●	●	◐	●	White	1	2	●	Champagne	3	6	●
Dutch Boy (K Mart) Fashion Fresh	14	L	●	○	◐	○	◐	○	●	White White	1	2	●	Honeycomb	2	5	●

1 *Too variable, sample to sample, to judge spatter.*   2 *This brand discontinued in the colors we tested.*

⑥ **Scrubbing.** The tougher a paint's dried surface, the more it can resist repeated cleanings.

⑦ **Water-resistance.** This measures how impervious to standing water a paint was. Important for surfaces likely to be wetted, such as tabletops, bathroom walls, and plant shelves.

⑧ **Blocking.** Blocking is a tendency for paint to stay tacky even after it has dried. More of a problem with latex paints, blocking makes some brands unsuitable for use on working surfaces such as bookshelves and tables.

⑨ **Hiding power.**

⑩ **Fading.** How well the individual colors hold up to the bleaching effects of sunlight. The lowest-scoring paints can fade even in indirect sun.

Properties specific to individual colors

Color name (Pink)	One-coat hiding ⑨	Two-coat hiding ⑨	Fading ⑩	Color name (Green)	One-coat hiding ⑨	Two-coat hiding ⑨	Fading ⑩	Color name (Blue)	One-coat hiding ⑨	Two-coat hiding ⑨	Fading ⑩	Color name (Yellow)	One-coat hiding ⑨	Two-coat hiding ⑨	Fading ⑩
Heathermist	2	6	◐	Green Whisper	2	6	◐	Country Blue	6	6	◐	Chrysanthemum	1	4	◐
Pixie Pink	2	5	◐	Pistachio	2	6	◐	Blue Magic	4	6	◐	Spring Tint	1	3	○
—	—	—	—	Mint Frost	2	6	◐	Crystal Blue	3	6	◐	Sunlight	2	4	◐
Sweet Clover	2	5	◐	Green Ice	2	5	○	Biscayne Blue	3	6	◐	Gin Fizz	1	3	○
Sweet Clover	2	6	○	Green Ice	2	6	◐	Biscayne Blue	5	6	◐	Gin Fizz	2	3	○
Stick Candy	1	5	●	Frosted Mint	4	6	●	Danish Blue	3	6	◐	Sugar Cookie	2	4	◐
Rose Mist	2	5	◐	Cool Eve	3	6	◐	Azure Foam	3	6	◐	Celestial Yellow	2	5	◐
Rose Mist	2	5	◐	Cool Eve	4	6	◐	Azure Foam	3	6	◐	Celestial Yellow	2	5	◐
Cotton Candy Pink	2	5	●	Iceberg Lettuce	3	6	●	Clearly Blue	4	6	●	Yellow Primrose	1	2	◐
Orange Blossom	1	3	○	Gossamer Green	2	6	○	Skyline Blue	4	6	○	Lemon Cream	1	2	○
Blush	2	6	●	Green Haze	4	6	●	Sky	4	6	●	Sunshine Yellow	2	5	●
Heathermist	1	3	●	Green Whisper	2	6	●	Country Blue	4	6	◐	Chrysanthemum	2	3	◐
Pixie Pink	2	6	◐	Pistachio	4	6	◐	Blue Magic	5	6	◐	Spring Tint	2	4	○
—	—	—	—	Orient Green	2	5	●	Hazy Blue	3	6	●	Lemon Yellow	2	5	◐
—	—	—	—	Mint Frost	1	4	●	Crystal Blue	4	6	●	Sunlight	1	3	○
—	—	—	—	Mint Frost	1	3	◐	Bellflower	2	6	●	Corn Yellow	1	2	◐

(continued)

Interior semigloss paints (continued)

Brand and model	Price	Gloss (1)	Brushing (2)	Leveling (3)	Sagging (4)	Spattering (5)	Resistance to: Scrubbing (6)	Water (7)	Blocking (8)	White: Color name	One-coat hiding (9)	Two-coat hiding (9)	Fading (10)	Gold: Color name	One-coat hiding (9)	Two-coat hiding (9)	Fading (10)
Dutch Boy (K Mart) The Fresh Look	18	L	●	◐	○	[1]	●	○	●	White White	1	3	●	Honeycomb	3	6	●
Fuller-O'Brien Double AA Series 214XX	21	M	●	◐	◐	○	◐	●	◐	White	1	4	●	Perfect Gold	6	6	◐
Fuller-O'Brien Ful-Flo Series 614XX	25	M	●	○	○	◐	●	●	●	White	2	4	◐	Perfect Gold	2	6	◐
Glidden Spred Enamel Series 3700	21	L	●	◐	◐	○	○	●	High Hiding White		1	3	●	Buttered Rum	2	6	○
Kelly-Moore Acry-Plex Series 1650	24	M	●	◐	○	●	●	●	White		1	3	●	Gold	4	6	●
Lucite Interior Semigloss Series 1600	19	L	●	●	●	●	●	○	●	White	1	3	●	Natural Beige	2	6	●
Magicolor Luster Plus Series 4211	18	L	●	●	●	●	●	●	Non-Yellowing White		2	4	●	Spice Beige	5	6	●
Pittsburgh Satinhide Series 88-00	21	L	●	○	◐	[1]	○	○	●	White	1	4	●	Cactus Flower	1	2	●
Pratt & Lambert Aqua Satin Series Z32372	26	L	●	●	○	●	●	○	●	One Coat White		2	4	◐	Wind Song	3	6	◐
Sears Easy Living Semi-Gloss Series 7100	15	L	●	●	●	●	●	○	●	Non-Yellowing White		2	5	●	Golden Harvest	2	6	●
Sears Easy Living for a Lifetime Series 7700	20	L	●	◐	●	●	●	●	○	●	Pure White	3	6	●	Golden Harvest	4	6	●
Sherwin-Williams Classic 99	21	L	●	◐	○	●	●	◐	●	Pure White	1	3	●	Caramel Corn	2	6	●
✓ Sherwin-Williams Superpaint	24	L	●	◐	●	◐	●	●	●	Pure White	2	6	●	Caramel Corn	4	6	●
Tru-Test Supreme E-Z Kare Series EZS	22	L	●	○	◐	●	●	○	◐	White	1	4	●	Sunny Mesa	2	6	●
Valspar Acrylic Series 42214	22	L	●	◐	●	●	●	○	●	White	2	5	●	Javelin	3	6	◐

Pink				Green				Blue				Yellow			
Color name	One-coat hiding	Two-coat hiding	Fading	Color name	One-coat hiding	Two-coat hiding	Fading	Color name	One-coat hiding	Two-coat hiding	Fading	Color name	One-coat hiding	Two-coat hiding	Fading
Pink Whisper	2	6	●	Mimosaceae	1	5	●	Bellflower	3	6	●	Yellow Bud	1	3	●
Pink Ruff	1	2	●	Marsh	5	6	●	Heidi	4	6	●	Marguerite	1	2	●
Pink Ruff	1	2	◐	Marsh	5	8	◐	Heidi	3	6	◐	Marguerite	1	2	◐
Sweet Clover	1	3	●	Green Ice	1	3	●	Biscayne Blue	2	5	●	Ginn Fizz	1	2	◐
Pink	2	6	●	Green	3	6	●	Blue	3	6	●	Yellow	1	3	◐
Rose Pearl	1	3	●	Spring Green	2	6	◐	Dove Blue	2	6	●	—	—	—	—
—	—	—	—	Mint Cooler	2	4	◐	Blue Horizon	2	6	●	Daffodil Yellow	1	4	◐
Stick Candy	1	2	●	Frosted Mint	2	5	◐	Danish Blue	2	6	●	Sugar Cookie	1	2	◐
Rose Mist	2	5	◐	Cool Eve	3	6	◐	Azure Foam	2	6	◐	Celestial Yellow	2	4	○
Apple Blossom	2	6	◐	Huckleberry Green	4	6	●	Federal Slate	3	6	●	Sunflower Yellow	2	4	◐
Apple Blossom	3	6	◐	Huckleberry Green	6	6	●	Federal Slate	5	6	●	Sunflower Yellow	2	6	●
Cotton Candy Pink	1	3	●	Iceberg Lettuce	2	6	●	Clearly Blue	3	6	●	Yellow Primrose	1	3	◐
Cotton Candy Pink	1	3	●	Iceberg Lettuce	4	6	●	Clearly Blue	3	6	●	Yellow Primrose	2	3	◐
Rose Quartz	1	4	●	Gossamer Green	2	6	●	Skyline Blue	1	4	●	Lemon Chiffon	1	2	○
Blush	2	5	◐	Green Haze	2	6	●	Sky	3	6	●	Sunshine Yellow	1	4	◐

Ratings
Exterior trim paints

Better ← → Worse

As published in *Consumer Reports,* September 1990.

Listed by types; within types, listed alphabetically. Brand-related properties apply to all colors tested for a brand. Color-related properties apply to a color. Dashes mean a suitable color wasn't available.

① Brand and model. If you can't find one of these paints, call the company.

② Price. In most cases, the manufacturer's approximate retail price per gallon. A * indicates the price we paid.

③ Gloss. The descriptions here are based on our measurements. Flat (**F**) is the dullest, followed by eggshell (**EG**), satin (**S**), semi-gloss (**SG**), gloss (**G**), and high gloss (**HG**).

④ Brushing ease. Alkyd formulations have improved over the years, but they're still harder to apply than latex paints.

⑤ Leveling. If you want a smooth finish, look to the alkyds.

⑥ Sagging. How well a paint resisted the tendency to run, drip, or sag like a curtain.

⑦ Adhesion. We applied paint to weathered panels that had been coated with a paint formulated to "chalk." When the panels had dried thoroughly, we scratched them and pressed tape over the scratch to see how much paint would pull away. But any of these paints should adhere to new surfaces or to properly prepared old paint.

① Brand and model	② Price	③ Gloss	④ Brushing ease	⑤ Leveling	⑥ Sagging	⑦ Adhesion	⑧ Blocking	White ⑨ Hiding	White ⑩ Color change	White ⑪ Chalking	White ⑫ Mildew	White ⑬ Dirt	Black ⑨ Hiding	Black ⑩ Color change	Black ⑫ Mildew	Black ⑬ Dirt
Alkyd (oil-based)																
Devoe All-Weather Gloss (Series IXX)	$32	SG	◒	◒	●	⊙	●	◒	●	◒	○	○	●	◒	◒	⊙
Devoe Velour Semi-Gloss (Series 29XX)	34	SG	○	◒	●	⊙	●	—	—	—	—	—	—	—	—	—
Dutch Boy Dirt Fighter Gloss (Series 1XX–2XX)	25	G	○	○	○	⊙	●	○	●	◒	○	○	●	◒	◒	◒
Fuller O'Brien Weather King (Series 660–XX)	28	G	○	○	○	⊙	●	◒	◒	●	◒	◒	●	◒	◒	●
Glidden Spred House Dura-Gloss Gelflow (Series 1900)	27*	G	○	○	○	⊙	●	○	●	◒	○	○	●	●	●	○
Moore's High Gloss (130)	26	SG	◒	◒	◒	⊙	●	⊙	◒	○	●	⊙	—	—	—	—
Moore's High Gloss Enamelized (Series 110)	24	[3]	◒	○	○	⊙	●	○	◒	○	◒	●	●	○	◒	●
Pittsburgh Sun-Proof Gloss (1 line)	31	G	○	○	○	⊙	●	◒	○	●	○	○	—	—	—	—
Pratt & Lambert Effecto High Gloss (E31172)[5]	36	HG	○	◒	●	⊙	●	◒	●	●	●	◒	●	◒	●	●
Pratt & Lambert Permalize Gloss (C34972)[5]	31	G	○	◒	◒	⊙	●	◒	●	●	○	◒	●	◒	●	●
Sears Best Weatherbeater Gloss (Series 4800)	23	SG	○	◒	◒	⊙	●	◒	◒	●	○	◒	—	—	—	—
Valspar Gloss (Series 2XX)	27	HG	○	○	◒	⊙	●	◒	◒	●	●	◒	●	◒	◒	◒

① Tends to erode rapidly when exposed to the weather. ② Brighter than most yellows. ③ Gloss too variable to rate. ④ Grayer than most blues. ⑤ Product number of white only.

8 Blocking. Paint, especially latex paint, can remain tacky long after it dries. The lower the score here, the more likely the paint will stick to things that touch it.

9 Hiding. A ◐ means the paint should cover almost any previous color in one coat. ◒ means the paint should cover in one coat if the old color doesn't contrast sharply with the new. Paints judged ○ should cover a similar color in one coat, a darker color in two. Paints judged ◓ or ● will require at least two coats to cover a similar color.

10 Color change. How individual colors stand up to the elements. The scores take into account both fading, yellowing (for whites), and loss of gloss.

11 Chalking. White paints with the lower scores chalked the most. They might be preferable in cities or suburbs, where dirt and pollution can quickly soil white paint.

12 Mildew. How well a paint resists the buildup of mildew. No paint will eliminate existing mildew; that requires washing.

13 Dirt. The faster a paint dries, and the smoother and more tack-free its surface, the less dirt it will attract.

Color-related properties

	Brown				Red				Green				Blue				Yellow			
	Hiding ⑨	Color change ⑩	Mildew ⑫	Dirt ⑬	Hiding ⑨	Color change ⑩	Mildew ⑫	Dirt ⑬	Hiding ⑨	Color change ⑩	Mildew ⑫	Dirt ⑬	Hiding ⑨	Color change ⑩	Mildew ⑫	Dirt ⑬	Hiding ⑨	Color change ⑩	Mildew ⑫	Dirt ⑬
	⊙	●	◒	⊙	—	—	—	—	⊙	●	◒	⊙	—	—	—	—	—	—	—	—
	—	—	—	—	●①	●	◒	◒	—	—	—	—	⊙	●	◒	◒	●②	●	●	◒
	⊙	●	◒	⊙	—	—	—	—	—	—	—	⊙	●	○	◒	—	—	—	—	—
	⊙	●	◒	⊙	—	—	—	⊙	●	◒	⊙	—	—	—	○	●	○	◒		
	⊙	●	◒	◒	—	—	—	⊙	●	◒	◒	—	—	—	—	—	—	—	—	—
	—	—	—	—	—	—	—	—	—	—	—	—	—	—	—	—	—	—	—	—
	⊙	●	◒	◒	—	—	—	⊙	◒	◒	◒	—	—	—	—	—	—	—	—	—
	⊙	●	○	◒	○①	⊙	◒	◒	⊙	●	○	◒	●④	●	◒	◒	●	●	◒	○
	⊙	●	◒	⊙	●①	○	●	◒	⊙	●	◒	⊙	⊙	●	◒	⊙	●②	●	●	◒
	⊙	●	○	◒	—	—	—	⊙	○	◒	◒	—	—	—	—	—	—	—	—	—
	⊙	●	○	◒	—	—	—	—	—	—	—	—	—	—	—	—	—	—	—	—
	⊙	●	◒	◒	●①	○	●	○	⊙	●	○	●	⊙	●	●	◒	●②	●	●	◒

(continued)

① Brand and model	② Price	③ Gloss	④ Brushing ease	⑤ Leveling	⑥ Sagging	⑦ Adhesion	⑧ Blocking	White ⑨ Hiding	⑩ Color change	⑪ Chalking	⑫ Mildew	⑬ Dirt	Black ⑨ Hiding	⑩ Color change	⑫ Mildew	⑬ Dirt
Latex (water-based)																
Ameritone Enamelized (W2500)	25	SG	⊙	●	●	⊙	●	●	◖	●	⊙	⊙	⊙	●	⊙	◖	⊙
Benjamin Moore Mooregard (Series 103)	20	EG	⊙	●	⊙	●	●	●	⊙	●	●	⊙	◖	⊙	●	⊙	●
Benjamin Moore Moorglo (Series 096)	22	S	⊙	●	●	●	●	⊙	◖	●	⊙	⊙	⊙	⊙	○	◖	○
Devoe Wonder Shield (Series 18XX)	28	SG	⊙	◖	◖	●	●	○	⊙	●	●	⊙	●	⊙	◖	⊙	⊙
Devoe Regency Satin (Series 19XX)	28	F	⊙	◖	◖	●	○	—	—	—	—	—	—	—	—	—	—
Dutch Boy Dirt Fighter Gloss (Series 19XX)	21	SG	⊙	●	⊙	●	◖	○	⊙	●	⊙	●	—	—	—	—
Dutch Boy Super Gloss (Series 74XX)	23	SG	⊙	●	⊙	●	●	◖	⊙	●	⊙	●	⊙	◖	⊙	◖
Dutch Boy Super Satin (Series 77XX)	22	EG	⊙	●	◖	●	●	○	⊙	●	⊙	◖	⊙	●	⊙	◖
Fuller O'Brien Versaflex Gloss (Series 615-XX)	30	G	⊙	●	⊙	●	●	—	—	—	—	—	—	—	—	—
Fuller O'Brien Weather King (Series 664-XX)	27	SG	⊙	◖	⊙	●	◖	⊙	⊙	●	⊙	⊙	⊙	⊙	⊙	◖
Glidden Spred House Dura-Gloss (Series 3900)	21*	SG	⊙	◖	◖	●	●	⊙	⊙	●	⊙	◖	⊙	⊙	⊙	◖
Glidden Spred House Dura-Satin (Series 2900)	21*	EG	⊙	◖	◖	●	●	⊙	⊙	●	⊙	◖	—	—	—	—
Lucite Enamel Gloss (Series 18XX)	18*	SG	⊙	●	◖	●	◖	◖	⊙	●	●	⊙	⊙	◖	⊙	◖
Lucite Satin (Series 22XX)	17*	S	⊙	●	⊙	◖	●	○	⊙	●	⊙	●	—	—	—	—
Pittsburgh Manor Hall Eggshell (79 Line)	30	F	⊙	◖	⊙	◖	●	○	⊙	●	⊙	⊙	⊙	⊙	◖	⊙
Pittsburgh Sun-Proof Semi-Gloss (78 Line)	26	SG	⊙	●	⊙	●	●	◖	⊙	●	⊙	⊙	⊙	⊙	⊙	⊙
Pratt & Lambert Aqua Royal Satin (Z3002) ⑤	29	SG	⊙	●	⊙	●	◖	⊙	⊙	●	⊙	◖	◖	⊙	●	◖
Sears Best Weatherbeater Satin (Series 5100)	20	EG	⊙	○	○	●	●	⊙	⊙	●	●	◖	⊙	◖	⊙	⊙
Sears Weatherbeater Premium Satin (Series 4700)	16	EG	⊙	●	⊙	●	◖	◖	⊙	●	●	○	⊙	◖	⊙	◖
Sears Weatherbeater Premium Semi-Gloss (Series 5000)	17	SG	⊙	●	◖	●	◖	⊙	⊙	●	●	●	⊙	⊙	⊙	◖

| | Brown | | | | Red | | | | Green | | | | Blue | | | | Yellow | | | |
	Hiding (9)	Color change (10)	Mildew (12)	Dirt (13)	Hiding (9)	Color change (10)	Mildew (12)	Dirt (13)	Hiding (9)	Color change (10)	Mildew (12)	Dirt (13)	Hiding (9)	Color change (10)	Mildew (12)	Dirt (13)	Hiding (9)	Color change (10)	Mildew (12)	Dirt (13)
	⊙	◐	⊙	⊙	—	—	—	—	—	—	—	—	—	—	—	●[2]	◐	⊙	●	
	⊙	◐	◐	⊙	—	—	—	—	—	—	—	⊙	⊙	⊙	○	●[2]	◐	◐	◐	
	⊙	○	⊙	○	○[1]	●	◐	◐	◐	⊙	◐	○	◐	◐	⊙	○	●[2]	◐	◐	◐
	⊙	○	◐	◐	—	—	—	—	●	◐	◐	○	—	—	—	—	—	—	—	—
	—	—	—	—	●[1]	◐	⊙	◐	—	—	—	—	●[1]	⊙	⊙	◐	●[2]	⊙	⊙	○
	⊙	○	⊙	◐	—	—	—	—	—	—	—	⊙	◐	◐	◐	○	◐	⊙	●	
	⊙	◐	⊙	◐	—	—	—	—	—	—	—	⊙	◐	◐	○	—	—	—	—	
	⊙	◐	⊙	◐	—	—	—	—	—	—	—	⊙	◐	⊙	◐	◐	⊙	⊙	◐	
	—	—	—	—	●[1]	●	⊙	○	—	—	—	⊙	⊙	⊙	◐	◐[2]	●	⊙	●	
	⊙	◐	⊙	◐	—	—	—	—	⊙	◐	⊙	◐	—	—	—	◐	◐	◐	●	
	⊙	⊙	⊙	⊙	—	—	—	—	⊙	⊙	⊙	◐	—	—	—	⊙	◐	⊙	◐	
	⊙	⊙	⊙	⊙	—	—	—	—	⊙	⊙	⊙	◐	—	—	—	⊙	○	⊙	○	
	⊙	◐	⊙	⊙	—	—	—	—	—	—	—	—	—	—	—	—	—	—	—	—
	—	—	—	—	—	—	—	—	—	—	—	—	—	—	—	—	—	—	—	—
	⊙	◐	⊙	◐	—	—	—	●	○	⊙	◐	⊙[4]	◐	⊙	●	◐	○	⊙	○	
	⊙	◐	⊙	◐	—	—	—	○	◐	⊙	◐	⊙[4]	⊙	⊙	◐	●	◐	⊙	●	
	⊙	◐	⊙	◐	—	—	●	◐	⊙	○	—	—	—	—	—	—	—	—		
	⊙	○	⊙	◐	—	—	⊙	◐	⊙	◐	●	○	⊙	◐	—	—	—			
	⊙	○	⊙	◐	—	—	◐	◐	⊙	◐	⊙[4]	◐	⊙	◐	—	—	—			
	⊙	●	⊙	◐	—	—	◐	◐	⊙	◐	⊙[4]	◐	⊙	○	—	—	—			

(continued)

Exterior trim paints (continued)

Brand and model	Price	Gloss	Brand-related properties							White				Black			
			Brushing ease	Leveling	Sagging	Adhesion	Blocking	Hiding	Color change	Chalking	Mildew	Dirt	Hiding	Color change	Mildew	Dirt
Sherwin-Williams A-100 Satin (A82 Series)	21	EG	⊙	●	⊙	●	◒	◒	●	●	●	●	◒	●	●	●
Sherwin-Williams Superpaint Gloss (A84 Series)	25	SG	⊙	●	⊙	●	●	●	⊙	●	●	●	●	—	—	—
Tru-Test Supreme Accent Color Semi-Gloss (AG Line)	24	SG	⊙	●	⊙	●	●	—	—	—	—	—	—	—	—	—
Tru-Test Supreme Weatherall Gloss (GHP Line)	24	SG	⊙	◒	◒	●	●	◒	●	●	●	●	●	◒	●	●
Valspar Semi-Gloss (Series 43XX)	24	S	⊙	●	◒	◒	●	◒	●	●	●	◒	●	○	●	◒

Names of colors tested

Alkyd paints

Devoe All-Weather Gloss: White, Black, Dark Brown, Forest Green. **Devoe Velour Semi-Gloss:** Hot Tango 2VR2A (red), Blue Saga 1BL11A, Dynamo 1BY26A (yellow). **Dutch Boy Dirt Fighter Gloss:** Gloss White, Black, Cocoa Brown, Triple Blue. **Fuller O'Brien Weather King:** White, Black, Sealskin (brown), Meadow Green, Sunray (yellow). **Glidden Spred House Dura-Gloss:** White, Black, Stratford Brown, Crylight Green. **Moore's High Gloss:** Brilliant White. **Moore's High Gloss Enamelized:** Outside White, Black, Tudor Brown, Chrome Green. **Pittsburgh Sun-Proof Gloss:** White, Bahama Brown, Colonial Red, Kentucky Green, Blue Mood, French Lacquer (yellow). **Pratt & Lambert Effecto:** High Hiding White, Black-Gloss, Leather Brown, Grenadier Red, Dublin Green, Postal Blue, Canary Yellow. **Pratt & Lambert Permalize:** High Hiding White, Black, London Brown, Glen Green. **Sears Best Weatherbeater:** White, Barcelona Brown. **Valspar:** Non-Chalking White, Black, Chocolate Brown, Cranberry (red), Forest Green, Blue (custom-mixed), Yellow (custom-mixed).

Latex paints

Ameritone: White, Black, Spanish Brown, Lustre Glow 1D39C (yellow). **Benjamin Moore Moorgard:** Brilliant White, Black, Tudor Brown, Blue 791, Yellow 321. **Benjamin Moore Moorglo:** White, Black, Tudor Brown, Tartan Red, Chrome Green, Blue 791, Yellow 321. **Devoe Wonder Shield:** White, Black, Dark Brown, Forest Green. **Devoe Regency Satin:** Hot Tango 2VR2A (red), Blue Saga 1BL11A, Dynamo 1BY26A (yellow). **Dutch Boy Dirt**

Brown				Red				Green				Blue				Yellow			
Hiding (9)	Color change (10)	Mildew (12)	Dirt (13)	Hiding (9)	Color change (10)	Mildew (12)	Dirt (13)	Hiding (9)	Color change (10)	Mildew (12)	Dirt (13)	Hiding (9)	Color change (10)	Mildew (12)	Dirt (13)	Hiding (9)	Color change (10)	Mildew (12)	Dirt (13)
⊙	◖	⊙	⊙	—	—	—	⊙	◖	⊙	⊙	⊙	—	—	—	●	◖	◖	◖	○
⊙	◖	⊙	◖	—	—	—	⊙	◖	⊙	○	—	—	—	—	◖	◖	◖	●	
⊙	◖	⊙	⊙	●[1]	⊙	◖	◖	○	⊙	○	⊙	○	⊙	◖	◖[2]	◖	⊙	●	
—	—	—	—	—	—	—	—	—	—	—	—	—	—	—	—	—	—	—	—
⊙	◖	⊙	◖	●[1]	◖	⊙	○	⊙	◖	⊙	◖	⊙	◖	⊙	◖	◖[2]	◖	⊙	◖

Fighter: White, Cocoa Brown, Triple Blue, Pawnee (yellow). **Dutch Boy Super Gloss:** White, Gloss Black, Cocoa Brown, Triple Blue. **Dutch Boy Super Satin:** White, Black, Cocoa Brown, Triple Blue, Pawnee (yellow). **Fuller O'Brien Versaflex Gloss:** Bright Red, National Blue, Sunshine (yellow). **Fuller O'Brien Weather King:** White, Black, Sealskin (brown), Meadow Green, Sunray (yellow). **Glidden Spred House Dura-Gloss:** White, Black, Stratford Brown, Crylight Green, Victorian Yellow. **Glidden Spred House Dura-Satin:** White, Stratford Brown, Crylight Green, Victorian Yellow. **Lucite Enamel Gloss:** White, Black, Bark Brown. **Lucite Satin:** White. **Pittsburgh Manor Hall:** Super White, Ebony Black, Bahama Brown, Kentucky Green, Blue Mood, French Lacquer (yellow). **Pittsburgh Sun-Proof:** White, Black, Bahama Brown, Kentucky Green, Blue Mood, French Lacquer (yellow). **Pratt & Lambert:** White, Black, London Brown, Glen Green. **Sears Best Weatherbeater:** White, Molten Black, Barcelona Brown, Mowbray Hunt Green, Tiber River Blue. **Sears House Shield:** White, Molten Black, Barcelona Brown, Desert Palm (green), Daring Indigo (blue). **Sears Weatherbeater Premium Satin:** White, Molten Black, Barcelona Brown, Azalea Leaf (green), Daring Indigo (blue). **Sears Weatherbeater Premium Semi-Gloss:** White, Molten Black, Barcelona Brown, Azalea Leaf (green), Daring Indigo (blue). **Sherwin-Williams A-100:** White, Tricorn Black, Chateau Brown, Mown Grass (green), Yellow Corn. **Sherwin-Williams Superpaint:** Super White, Chateau Brown, Mown Grass (green), Yellow Corn. **Tru-Test Supreme:** Tru-Brown, Tru-Red, Tru-Green, Tru-Blue, Tru-Yellow. **Tru-Test Supreme Weatherall:** White, Black. **Valspar:** White, Black, Chocolate Brown, Cranberry (red), Forest Green, Dominion Blue, Oleo Yellow.

Ratings
Stepladders

These listings are taken from a September 1990 report.

Listed by type. Within types, listed by height in order of estimated quality, based primarily on safety and on convenience as judged by a panel of users.

❶ Brand and model. The fiberglass/aluminum models are popular with contractors. The others are sold primarily to do-it-yourselfers. Fiberglass/aluminum and wood ladders can safely be used anywhere. Aluminum ladders should not be used outdoors with electric tools because the ladders can conduct electricity. If you can't find a model, call the company.

❷ Price. The average price we paid. A + indicates that shipping was extra.

❸ Working load. Manufacturer's claims for the maximum weight a ladder should bear, as determined in standard industry tests. These figures correspond to the weight limits in the duty rating displayed on each ladder. For safety's sake, your weight plus that of the materials you bring onto the ladder with you should never exceed the working load. A 200- or 225-pound working load is enough for most people.

❶ Brand and model	Price ❷	Weight	Working load ❸	Resistance to — Swaying ❹	Walking ❺	Tipping ❻	Ease of — Opening & closing ❼	Carrying ❽	Moving ❾	Advantages	Disadvantages	Comments
Fiberglass/aluminum, 6-ft.												
Keller 976	$ 92	21 lb.	300 lb.	●	○	○	●	○	◐	A,B	—	A,E
Werner 7206	168	23	300	●	○	◐	◐	○	○	A,E	h	A,E
Aluminum, 6-ft.												
Werner 376	120	14	250	◐	○	◐	●	◐	◐	A,B,C,D,E,F	—	A,C
Keller 916	63	17	250	◐	◐	◐	◐	○	○	B,C,D	—	A
Keller Greenline 926	53	12	225	◐	○	◐	◐	◐	◐	B,C,D	—	A
Keller Redline 936	48	11	200	◐	○	◐	○	◐	◐	B,C,D	—	A
Sears Craftsman 42176	50+	14	225	◐	◐	○	◐	◐	◐	A,D,	c,h	A,G,H
Werner 366	74	12	225	○	◐	◐	◐	◐	◐	A,B,C,F	—	A
Sears Craftsman 42216	40	11	200	◐	◐	◐	○	◐	◐	D	f,h	H
Werner 356	60	12	200	○	◐	◐	●	◐	◐	B,F	—	A
Aluminum, 8-ft.												
Keller Greenline 928	93	17	225	○	◐	◐	●	○	◐	C,D,	h	A
Sears Craftsman 42178	90	18	225	◐	◐	◐	◐	○	◐	A,D	h	A,G,H

Specifications and Features
All 6-ft. models: Let you safely stand no more than 3 ft. 7 in. to 3 ft. 10 in. from ground.
All 8-ft. models: Let you safely stand no more than 5 ft. 3 in. to 5 ft. 9 in. from ground.
Except as noted, all models have: Grooves or traction treads on steps to deter slipping.
All wood models have: Steel tie rods under steps for structural integrity.
All aluminum and fiberglass/aluminum models have: Pad at bottom of each leg to protect floors and deter slipping.

Key to Advantages
A—Space between side rails reduces pinching hazard.
B—Spreaders, which hold ladder open, unlikely to pinch fingers.
C—Pail shelf has rounded corners. For 6-ft. models, that design judged less hazardous than sharp corners on others; for 8-ft. models, rounded corners matter less, since shelf is 6½ ft. from floor.
D—Pail shelf folds automatically when ladder closes.
E—Spreaders are protected from damage because they are inside side rails.
F—Has clips to keep folded ladder closed (but clips are easily lost).

④ **Resistance to swaying.** A key judgment from our panelists. It shows the stability of a ladder as you climb, shifting your weight from step to step, or perch on the highest step you can safely occupy.

⑤ **Resistance to "walking."** This panel-test judgment shows which ladders are most likely to chatter along the floor as you shift your weight from step to step. A ladder that "walks" too much can scuff the floor, upset a can of paint, or even make you fall.

⑥ **Resistance to tipping.** Our panelists gauged tippiness by putting one foot on the ladder's lowest step and grabbing the ladder with one hand while carrying a small but bulky load

in the other. The wider the angle between an open ladder's front and back legs, the less likely the ladder is to tip.

⑦ **Ease of opening and closing.** The panelists opened and closed each ladder several times. Balky hinges, sticky pivots, and such lowered a ladder's score.

⑧ **Ease of carrying.** A difference of a pound or two didn't seem to matter to our panelists, who carried the ladders closed.

⑨ **Ease of moving.** How easily you can pick up an opened ladder and move it a few feet. Light weight, a comfortable handhold, and proper balance were assets our panelists looked for.

| | | | | Resistance to | | | Ease of | | | | | |
① Brand and model	Price ②	Weight	Working load ③	Swaying ④	Walking ⑤	Tipping ⑥	Opening & closing ⑦	Carrying ⑧	Moving ⑨	Advantages	Disadvantages	Comments
Wood, 6-ft.												
Werner W356	29	20	200	○	○	◐	●	○	○	A,B,D	d	—
Werner W366	58	22	225	○	○	○	◐	◐	◐	A,B,D,F	—	A
Putnam Peerless	59	22	225	◐	○	◐	●	○	○	A,D	h	D
Sears 40306	20	17	200	◐	◐	◐	●	○	◐	A,D	b	B,H
Lynn No. 76 Supertred SUS06	58	21	225	◐	○	◐	◐	○	○	A,B,D	a,g	—
Werner W336	38	17	200	◐	○	○	◐	◐	○	A,B,D,F	d,g	—
Lynn No. 75 Patriot PA506	46	19	200	◐	○	○	◐	◐	○	A,B	a,g	D
Archbold 16006	38	21	225	◐	○	○	○	○	○	A,B,D	b,d,g	—
Keller W2-6	27	21	225	◐	◐	○	○	○	○	A,B,D	g	—
Keller W-6	32	18	200	◐	◐	○	○	◐	○	A,B,D	a,d,g	—
Putnam Durable	48	20	200	◐	○	○	○	○	◐	A,D	h	D
Archbold 14006	28	19	200	●	○	○	○	◐	○	A,B,D,	a,b,g	F
Wood, 8-ft.												
Werner W368	78	30	225	◐	○	◐	●	◐	◐	A,B,D	i	A
Keller W2-8	56	30	225	◐	○	○	◐	◐	◐	A,B,D	d,e	—

Key to Disadvantages

a—Wobbled noticeably without load.
b—Leg bottoms were not horizontal to floor and could damage it.
c—Failed American National Standards Institute test for step-bending.
d—Tested samples had splinters as received.
e—Pail shelf broke when ladder was closing.
f—Pail shelf judged likely to strike you in the face when folding ladder.
g—Pail shelf could strike you in the face when folding ladder, a slight hazard.
h—Spreaders could pinch or squeeze fingers when you're closing the ladder.
i—Truss block, which supports steps, fell out when we tested ladder's ability to withstand severe load.

Key to Comments

A—Top plate or pail shelf accommodates tools.
B—Lacks spreaders; pail shelf keeps ladder open.
C—Has "H"-shaped spreaders with handle; allows ladder to be opened and closed with one hand.
D—Has smooth (not grooved) step surfaces, judged comfortable for standing.
E—Lacks pail shelf.
F—Tie rods under steps judged likely to need tightening more often than on most wood models.
G—Has 5-yr. warranty.
H—Discontinued but may still be available in some Sears retail stores. Replacements as follows: **40116** by **40306**, $20; **42386** by **42176**, $50; **42156** by **42216**, $40; **42388** by **42178**, $80.

Ratings
Finishing sanders

From a September 1990 report.

Listed in order of estimated quality. Except where separated by a bold rule, closely ranked models differed little in quality.

❶ Brand and model. We tested moderately priced models for occasional use and more expensive professional models. If you can't find a model, call the company.

❷ Price. The manufacturer's suggested retail price. Discounts are common.

❸ Type. The **palm** sanders, designed to be guided with one hand, are better suited for experienced users. The **two-handled** models are better for novices because they are less likely to sand depressions into the wood.

❹ Sheet size. The part of a standard 9 × 11-inch sheet of sandpaper each sander holds: **quarter-sheet,** about 4½ × 5½ inches; **one-third-sheet,** about 3½ × 9 inches; **half-sheet,** about 4½ × 11 inches. The **one-sixth-sheet** model takes a sheet about 3½ × 4½ inches. One-third-sheet models are a good in-between size for all-around work; larger models may not perform well in tight quarters, while the smaller ones have to be moved continuously to sand uniformly.

❺ Weight. In ounces, including the cord. Weight only matters when you're working overhead or on a vertical surface; you have to support the sander and hold it against the work.

❻ Sanding speed. Shows how rapidly each sander removed wood from pine boards using 100-grit paper. (Sanders that can be switched between orbital and back-and-forth motion

❶ Brand and model	❷ Price	❸ Type	❹ Sheet size	❺ Weight	❻ Sanding speed	❼ Clamp quality	❽ Handling	❾ Evenness	❿ Noise	⓫ Vibration	⓬ Pad contact	⓭ Cord length	Advantages	Disadvantages	Comments
Black & Decker 4011	$ 79	Palm	¼	47 oz.	◒	●	◒	◒	○	●	◒	96 in.	A,B,D,G,I,J,K	—	A,D
Black & Decker 7458	122	2-handled	½	64	◒	●	○	◒	○	○	●	72	A,D,E,F,K	c,p,s,u	B,F
Hitachi SV12SA	80	Palm	¼	47	◖	●	○	○	○	●	◒	100	F,J	—	A
Makita BO4550	86	Palm	¼	37	◖	◒	●	○	○	◒	◒	74	A,B,D,F,J	—	B,H
Sears Craftsman 11602	33	Palm	¼	40	◖	◒	○	○	◒	○	●	96	F,H	h	A,B
Skil 7582	121	2-handled	⅓	71	●	○	◒	◒	◒	◒	◒	96	J	a,g,i,u	B,C
Black & Decker 7448	47	2-handled	⅓	37	◖	◖	◒	◒	◒	○	◒	72	I	a,u,v	B,F
Ryobi S500A	79	Palm	⅙	42	◖	◒	◒	◒	◒	◒	○	79	F,J	a,m	—
Makita 9035	112	2-handled	⅓	56	◖	●	◒	○	◒	◒	◒	82	I,J	a,l,u,v	B,H

Specifications and Features
All: Run on 120 volts and are double insulated. *Except as noted all:* ● Use only orbital motion. ● Lack provision for dust collection. ● Can sand to within ⅛-in. of perpendicular edge on at least 1 side. ● Have rocker or toggle On/Off switch that's easy to operate but poorly marked. ● Have easily used finger-operated clamps to hold paper; clamps tend to pull paper taut as they close. ● Come with good instruction manual. ● Have reasonably limp power cord, which doesn't interfere with work. ● Judged suitable for use with self-adhesive sandpaper. ● Can be used on wood, metal, and plastics, or to remove paint. ● Have 1-yr. warranty against defects in noncommercial use.

Key to Advantages
A—Has through-the-paper dust collection; requires punching holes in paper with tool provided.
B—Paper punch cuts all holes at once, with paper on sander.
C—Has dust-collecting skirt around base.
D—Has dust-collecting bag.
E—Variable speed.
F—Sands to within less than ⅛-in. of perpendicular edge of front and sides.
G—User can replace motor brushes without disassembling sander.
H—On/Off switch has better markings than most.
I—Power cord limper than most; less likely to interfere with sander's motion across work.
J—Judged more suitable than most repairs by user.
K—Instructions better than most.
L—Has switch to convert sander from orbital to straight-line motion.

Key to Disadvantages
a—Uses ⅓ or ⅙ of standard sandpaper sheet; often requires measuring to cut sheet to proper size.

were tested in their orbital mode.) The best sander was nearly three times faster than the slowest.

7 Clamp quality. A major point of convenience with any finishing sander. The best clamps open and close easily, have a wide opening, and keep the paper taut and flat against the sander's base.

8 Handling. How easy it was to guide the sander across a board. In our tests, some sanders hopped, rotated, or resisted changes in direction.

9 Evenness. It's difficult to sand wood with a wide grain, such as fir. The light parts of the wood are softer than the dark parts and wear away faster. The better models sanded fir plywood without leaving valleys in the softer parts. Sanders with lower scores left noticeable valleys.

10 Noise. The better the score, the quieter the sander. The loudest exceeded 90 decibels, a level that we think requires hearing protection.

11 Vibration. Our assessment of each sander's ability to minimize the vibration you'll feel when you use the tool. The worst will leave your hands tingling.

12 Pad contact. How evenly each sander's pad contacted the surface being sanded. A low score means that a large part of the pad did not touch the surface, increasing the possibility of sanding unevenly. In actual practice, if the sandpaper is tight and the sander kept in constant motion, pad contact won't matter much.

13 Cord length. The longer the power cord the better, provided the cord isn't so stiff it impedes your ability to move the sander across the wood. Models with particularly good or bad cords are noted in the Advantages and Disadvantages.

Brand and model	Price	Type	Sheet size	Weight	Sanding speed	Clamp quality	Handling	Evenness	Noise	Vibration	Pad contact	Cord length	Advantages	Disadvantages	Comments
Skil 7382	69	2-handled	⅓	55	●	◐	○	○	○	◐	◐	60	C,D	a,k,l,q,r,s,v	B
Milwaukee 6016	84	Palm	¼	46	○	◐	○	●	◐	○	●	105	F,H	n,s,v	B,E
Skil 7575	70	Palm	¼	43	◐	◐	○	◐	○	◐	●	96	F,I,J	d,u	B
Skil 7576	78	Palm	¼	43	◐	◐	○	◐	○	◐	◐	96	A,B,D,I,J	b,d,u	B
Ryobi S600	90	Palm	¼	46	●	○	◉	●	◐	○	◉	80	J	k,v	G
Sears Craftsman 11611	23	2-handled	⅓	50	○	●	○	○	◐	○	●	72	F,H	a,e,n,s	B
Ryobi LS35	98	2-handled	⅓	54	◐	◐	○	○	◐	◐	●	80	J	a,g,m,p,w	—
Sears Craftsman 11616	67	2-handled	½	83	○	●	◐	◉	●	●	○	122	C,D,H,L	e,i,k,l,n,s,t,u	B,D
Porter-Cable 330	97	Palm	¼	61	●	◐	◐	◐	◐	◉	◉	82	I,J	d,f,i,k,v	B
Sears Craftsman 11613	55	2-handled	⅓	67	◐	●	◐	◐	●	●	●	72	C,D,F,H,L	a,e,i,l,n,p,s,t,u	B,D

b—Punch for dust-collecting system doesn't perforate paper cleanly.
c—Punch cuts only 1 hole at a time in sandpaper.
d—Harder to load paper than most.
e—Clamps do not tend to tighten paper.
f—Clamps require tool to operate easily.
g—Clamps require strong fingers to operate.
h—Clamps protrude from base far enough to scratch adjacent vertical edge of work.
i—Instructions worse than most.
j—Sanded no closer than ⅛ to ⅜-in. from adjacent perpendicular edge.
k—Dust-collecting skirt must be removed to get sander close to adjacent perpendicular edge.
l—On/Off trigger switch requires two operations to lock sander on. Judged hard for left-handed persons to use.

m—On/Off switch harder to operate than most.
n—On/Off switch more likely than most to be operated by accident.
o—Front handle judged too small for people with large hands to control easily.
p—Front handle judged too large for people with small hands to control easily.
q—Front handle judged too low for good control when using sander at arm's length.
r—Main handle judged too low for good control when using sander at arm's length.
s—Power cord stiffer than most; could interfere with sander's motion.
t—Dust-collecting bag not securely fastened to sander.
u—Judged unsuitable for use with self-adhesive paper.
v—Instructions judged only fair.

Finishing sanders (continued)

w—Levers to switch from orbital to straight-line sanding confusingly marked.

Key to Comments
A—Round base available as option (not tested).
B—Instructions don't recommend users replace motor brushes.
C—Instructions cover more than one sander.
D—Optional dust-collecting system allows sander to be connected to vacuum cleaner; can be useful for people

sensitive to wood dust, but may make sander awkward to use.
E—Has unlimited warranty against defects in noncommercial use.
F—Has 2-yr. warranty against defects.
G—Instructions say to use sander on wood only.
H—According to instructions, motor brushes are self-limiting; when brushes wear out, the motor won't run, preventing major damage to motor.

Index